OTHER WORLDS

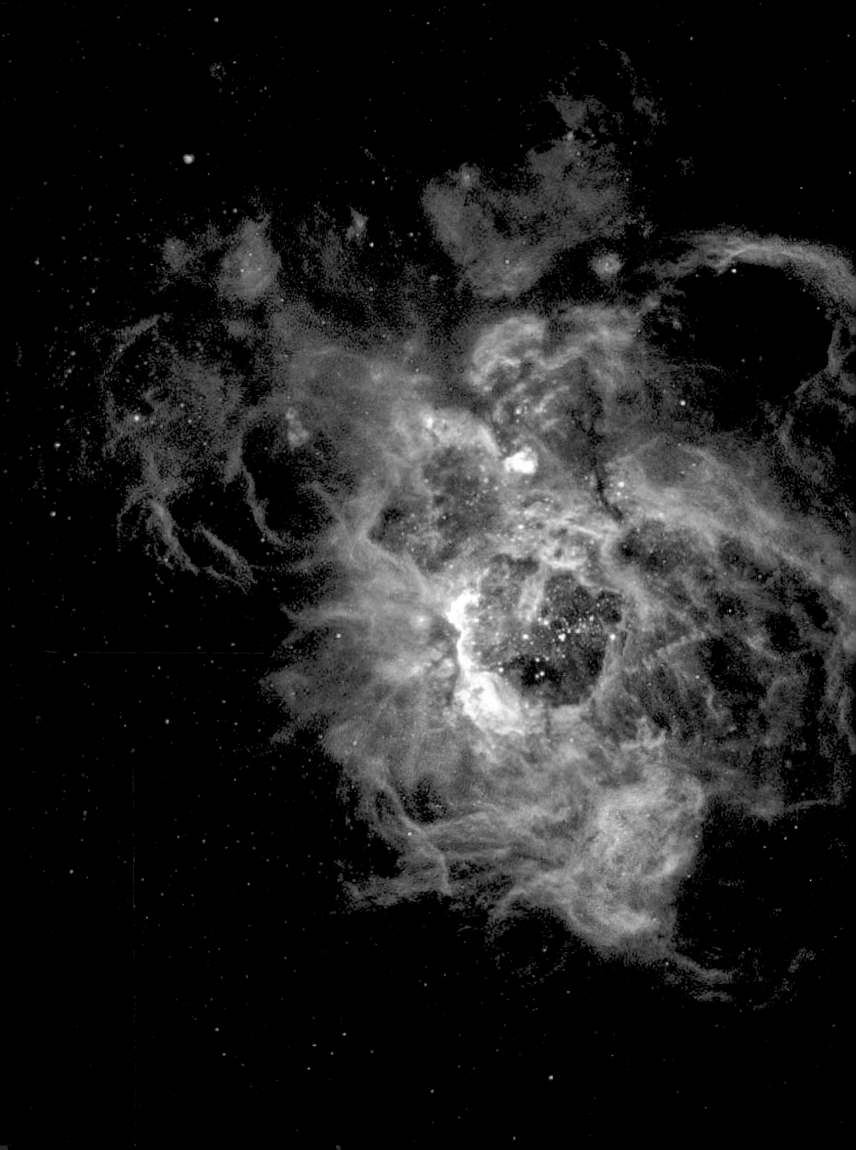

OTHER
WORLDS

IMAGES OF THE COSMOS
FROM EARTH AND SPACE

BY JAMES TREFIL

FOREWORD BY DAVID H. LEVY

NATIONAL
GEOGRAPHIC

WASHINGTON, D.C.

CONTENTS

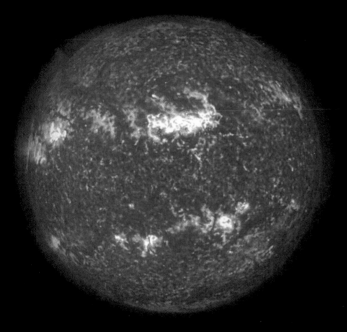

BIRTH OF THE SOLAR SYSTEM

THE INNER PLANETS

[PREVIOUS PAGES] One image of creation, the vast star-birthing nebula known as NGC 604 spans nearly 1,500 light-years. At its glowing heart more than 200 hot, massive stars heat the surrounding gas, causing it to fluoresce.

[ABOVE, LEFT TO RIGHT] The sun, Mars, Jupiter, and M100, one of the brightest galaxies of the Virgo cluster.

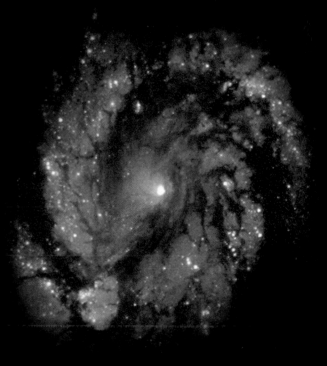

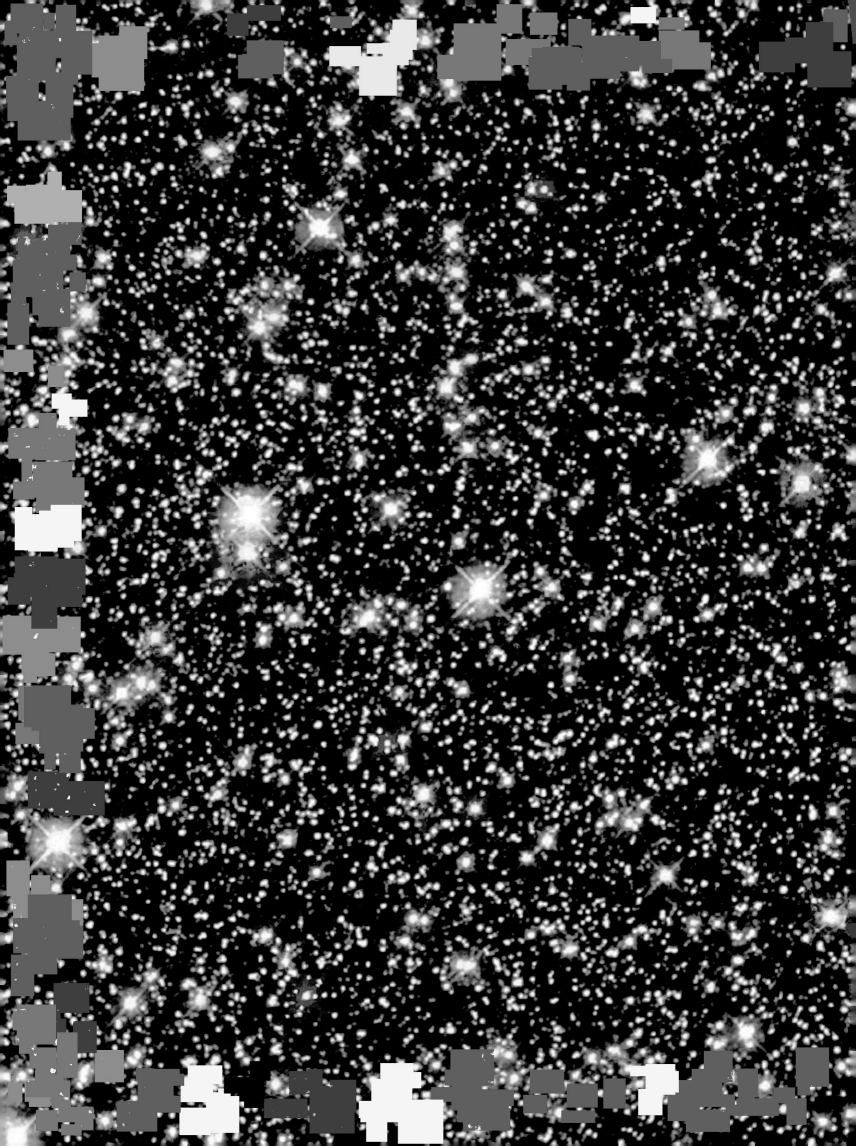

OF SPACE AND TIME

BY DAVID H. LEVY

WITH ITS MOON AND PLANETS AND MULTITUDE OF STARS, THE NIGHT SKY FREES our imaginations to wonder. How many distant stars and galaxies are out there? How many other worlds? And are we alone, with Earth the only planet blessed with life?

My lifelong career as a serious stargazer began in the summer of 1960, when I was 12 years old, and it was helped on its course by the gift of a wristwatch. As my eyes followed the movement of the second, minute, and hour hands of that watch, my mind left Earth to imagine the greater timepiece of the solar system, with planets racing around the sun in clockwork precision. And as I began to grasp the awesome infinity of space and time that is the cosmos, I thought about how tired that little watch would have been if it had marked time since the universe began, its hands whirling around and around for perhaps ten billion years.

Watches and telescopes are vital to astronomers. For one thing, modern science sees space and time as dual manifestations of the same thing. Clocks measure time and time measures space, especially astronomical space. We calculate distances between stars and galaxies in terms of light-years—the distance light travels in a year (about six trillion miles). Because light takes time to travel, watching the night sky also is a voyage backward in time. The farther a star is from us, the longer its light takes to reach us; therefore, whatever we see today actually happened long, long ago.

My first telescope was a modest one, packing a mere 3.5-inch mirror. But to me it was a powerhouse mightier even than the famous 200-inch behemoth atop California's Palomar Mountain—for it was accessible to *me*. Just as Palomar helped open the universe to astronomers, that small scope enabled a boy on the edge of teenhood to enjoy and celebrate the night sky.

I remember the anticipation I felt as I assembled it. That evening, the deepening sky quickly filled with innumerable twinkling points of light, all of them beckoning for a closer look. I had

LURE OF THE NIGHT SKY

Gemlike members of the Sagittarius star cloud—a relatively dust-free region toward the center of our galaxy, the Milky Way—exhibit vivid colors and a range of intensities in this image from the Hubble Space Telescope. Some of these stars rank among the galaxy's oldest; by studying them, scientists hope to learn more about how the Milky Way evolved.

no star chart back then, so I just chose one of the brighter objects, up above the trees to the south. I focused the eyepiece and watched, spellbound, as a marvelously striped ball surrounded by four "stars" took shape. Even at age 12, I knew that I had stumbled upon the solar system's largest planet, Jupiter, accompanied by its four largest and most visible moons.

Night after night I would gaze again on Jupiter, watching those moons march around and around, keeping time much like the hands of my wristwatch. I didn't know it, but back in those early days of personal discovery, neither my telescope nor Palomar's was able to detect a different sort of satellite that hovered near that planet. For while Jupiter's biggest moons take only days or weeks to complete a revolution, this one circuited the planet just once every two years! And while all of Jupiter's 16 moons, large and small, move in stable circular orbits, this one worked its way around the planet in narrow and greatly elongated loops.

Some 30 years later, in 1993, I was searching the sky for asteroids and comets with two other veteran stargazers: Gene Shoemaker, the astronomer and geologist, and his wife Carolyn, also an astronomer. I took two photographs of a particular field of sky, which happened to include Jupiter. As I carefully guided the telescope, I recalled my first glimpse of this planet so many years before. A lot had happened since 1960, yet I was still watching the sky each night, and Jupiter's moons were still orbiting the planet exactly as they always had.

At least that's what we thought—until Carolyn scanned two Jupiter films and noticed what seemed to be a bar of light, with several tails. She said it looked like "a squashed comet," and in fact that's just what it was: a comet that was being shattered by Jupiter's immense gravity. Thus did Comet Shoemaker-Levy 9, that Jovian "moon" with the greatly elongated orbit, announce itself to our world. Even then, however, we had no idea that our discovery was just the overture to a thrilling finale that would play itself out a year later.

Celestial mechanicians soon calculated that our shattered comet was in a kamikaze orbit, destined to fragment further and eventually collide with Jupiter. That finally happened in July 1994 when, over the course of a single week, all those chunks of Shoemaker-Levy 9 crashed into Jupiter, one every eight hours or so. It was a stunning show, and it proved that comets play an important role. This one dropped tons of water on Jupiter, just as ancient cometary impacts must have deposited water, carbon, nitrogen—the building blocks of life—on Earth's surface billions of years ago. Much more recently ("only" 65 million years ago, that is) a mighty comet or asteroid smashed into Earth, wiping out most living things, including the dinosaurs.

INNUMERABLE EVENTS OCCUR IN THE SKY ABOVE US, BUT THERE ARE TWO overpowering basics. The first is motion: Planets revolve around the sun, and moons revolve around planets. We see evidence for this weekly and even nightly, as our moon changes phase and, over the longer term, as we chart the progress of planets around the sun. The second basic truth is that moving bodies may collide. This does not happen nightly or even weekly, but it did happen with Shoemaker-Levy 9. For me, that experience epitomizes the excitement and hope that drive all stargazers. We watch because we want to see and to learn. My own initial glimpse of Jupiter in 1960 eventually led me on a lifelong pursuit—which has culminated, so far, in the discovery of 21 different comets, including the one that smashed into Jupiter.

What's it like to be a skywatcher? For me, the nights of the past 38 years have been defined by phases of the moon and exigencies of the weather. On clear and moonless nights—the best nights, from my perspective—I may set up my telescope and search all night long for comets. If the moon appears in early evening, I might retreat, setting my alarm clock for the early morning hours when the moon will be gone and I can return to spend an hour or two with the stars. Comet hunting is a quiet occupation. In those dark hours before dawn, I often find myself alone with my telescope, the stars, and an immense sense of tranquillity.

Such a nocturnal lifestyle demands some adjustments. My friends, for example, all know not to phone me before early afternoon! It's not always easy being a night person, but there are rewards. Despite all the time I've spent watching the night sky, I have never lost my amazement at what can happen up there. This love of astronomy did not come from college or any formal instruction. In fact, I've never taken a course in astronomy. I've learned all I know—and don't know—by reading, observing, and being lucky enough to work with two of the best astronomers this century has seen, Gene Shoemaker and Clyde Tombaugh.

Tombaugh, the only person of this century to have discovered a major planet in our solar system, knew the sky as well as you or I know our houses. He discovered Pluto, the outermost planet, back in 1930 during an exhaustive search that involved examining some 90 million points of light, one by one. I still admire the way he got to be the chief "planet hunter" at Arizona's Lowell Observatory: He had been studying Mars and Saturn through a home-built telescope, and one day decided to send Lowell's director some rough pencil sketches of those planets. All he expected was a critique of his art. Instead, he ended up with a job offer that eventually would lead to the discovery of Pluto. In many ways, Clyde's life typified what astronomy is all about: becoming one with the sky through hours of patient observation and study.

Gene Shoemaker also was a consummate observer, one who became convinced that comet and asteroid impacts have had major effects on the course of life on Earth. He began an organized search of the sky, not for planets, as Clyde did, but for asteroids and comets. He and Carolyn together took more than 26,000 photographs of the night sky. I joined their program in 1989, and of the 21 comets I've found, 13 were first sighted through the 18-inch diameter telescope atop Palomar Mountain, which I shared with the Shoemakers. Sadly, Gene was killed in a car accident in 1997. But Carolyn and my wife, Wendee, and I continue to search for comets, relying usually on two telescopes set up near our Arizona homes.

In addition to years and years of observation, reading has bolstered my perception of the night sky. Just before I got that first telescope back in 1960, my cousin gave me an astronomy book, and as I read each section I felt as though I had embarked on an actual journey of discovery, pushing from the innermost planets to the outer "gas giants," then beyond to our galaxy, the Milky Way, to other stars adrift in other galaxies, and finally to the farthest regions of the universe. My life has been a voyage as well. Along the way I've met other travelers, such as Tombaugh and the Shoemakers, and sailed awhile with them. I've also been aided by the knowledge encapsulated in numerous articles and books.

Among those articles were at least a dozen from NATIONAL GEOGRAPHIC. One, which described tracking a total eclipse of the sun back in 1947, was written by F. Barrows Colton, an editor at the Magazine during the 1940s and 1950s, and the father of a childhood friend of mine. Since then, of course, the GEOGRAPHIC has chronicled many astronomical and space events, including the especially memorable one dedicated to John Glenn's pioneering orbital flight in 1962.

And here you have before you the latest GEOGRAPHIC effort, filled with new information and truly spectacular images gathered from numerous observatories and space missions, including the Hubble Space Telescope. Now you can set out on your own incredible journey, exploring each of the planets and their moons, the asteroids and comets, then moving on to the vast home galaxy of which we are such an infinitesimal part. *Other Worlds* also examines the realm of supernovas, pulsars, and black holes—the mysterious end-stages in a star's life—as well as quasars, those incredibly powerful energy sources that exist at the edge of the visible universe. I remember when regular, pulsing signals were first reported from space, in 1968. Thought by some to be evidence for extraterrestrial life, these signals were first dubbed LGMs, for "little green men." In time, they were explained as radiations of a spinning pulsar that, like a lighthouse beacon, seems to blink on and off—to "pulse"—as it rotates.

I first learned of quasars in the 1960s, when astronomers discovered a mysterious source of radio waves coming from Virgo. It was so distant that no one knew what it was; it soon became labelled a "quasi-stellar object," quasar for short. To date, astronomers have found hundreds of other quasars, now considered to be the blazing cores of far-off galaxies. They remain among the most distant objects in space and they possess unimaginable amounts

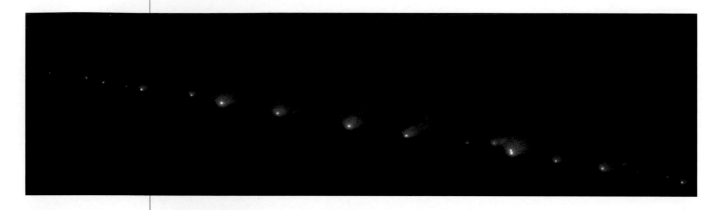

DEATH OF A COMET

Racing toward Jupiter, doomed Comet Shoemaker-Levy 9 broke into 21 icy chunks by May 1994, creating a stream of glowing dots across more than 700,000 miles of space (above).

Each fragment collided with the planet two months later, in fiery displays (opposite) that raised plumes thousands of miles high and left dark impact scars on the planet's face.

of energy. We're lucky that our galaxy isn't centered on a quasar, for if it were, our nights would be continually lit by a distant ball as bright as the full moon.

Although the main ports of call on my voyage through the night sky have been comets, I still gaze at and even draw the moon's craters, and follow the behavioral antics of Cepheid variables, stars that vary in brightness cyclically. I also count myself extremely fortunate to have witnessed, back in 1975, the exploding star known as V1500 Cygni.

To astronomers, there are few things as exciting as an unexpected astronomical event, and exploding stars top the list. V1500 Cygni was not destroying itself; rather, it was having a tantrum and blowing off some of its outer atmosphere. Twelve years later, I saw an even rarer phenomenon: a massive star in the Large Magellanic Cloud (LMC), the closest galaxy to the Milky Way, had run out of fuel, collapsed on itself, and then blown itself apart. For a few weeks this shattered star—technically called a supernova—emitted more energy than all the stars in the LMC *put together*, reaching across the 160,000 light-years that separate it from us. It was the brightest supernova in nearly four centuries.

In addition to witnessing such grand and distant displays, I've often enjoyed the aurora borealis, especially when I lived in Canada. Auroras occur just a few tens of miles overhead, so my personal journey has encompassed the depth of the night sky—from local happenings in our atmosphere to the farthest galaxies and quasars. *Other Worlds* enables you to undertake a journey that's similarly varied, one that touches on the diverse worlds of our own solar system and goes on to stars far beyond our sun, to other wonders at the edge of the cosmos. Perhaps you'll recall, as you turn these pages, the continuously moving hands of *your* first watch, faithfully ticking off the passage of time just as the universe has done for billions of years. ●

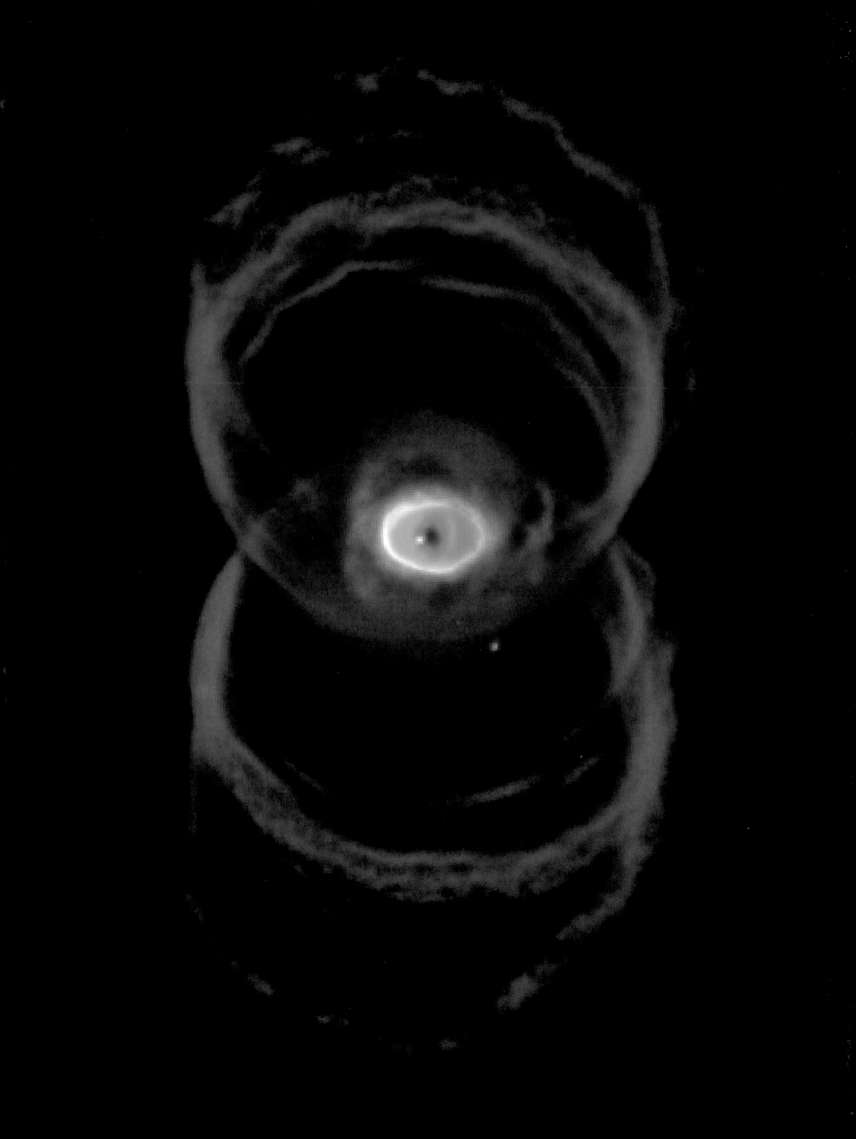

A QUESTION OF SCALE

THE UNIVERSE MAY BE BIG, BUT OVER THE LAST TWO MILLENNIA HUMAN BEINGS have learned an awful lot about it. And as our ideas have changed, so too have our notions of where we fit into the grand scheme of things.

I like to compare the universe that modern science has unveiled to a *matryoshka*—one of those Russian dolls that opens up to reveal a smaller doll inside, which in turn contains a still smaller doll, and the process goes on until you get to the end, the smallest doll in the set.

Our exploration of the universe has been a little bit like opening a matryoshka in reverse. Starting with our own planet, which for all its majesty is still the smallest doll—we have worked our way outward to the solar system (the next biggest doll), the stars of our own Milky Way (the next biggest), all the way to the relationships between the galaxies themselves (the largest doll we can imagine, so far). Each step in this process has required the best minds the human race could produce, not to mention previously undreamed of advances in technology. From the crude telescopes that first detected Saturn's rings to the Cassini spacecraft that is heading toward those rings as I write these words, there stretches a continuous line of human beings, often toiling in obscurity, without whom our current, magnificent vision of the universe would never have come to be.

Before we get into their story and the results of their vision, however, it probably would be a good idea to orient ourselves to our universe—to get some idea about the sizes of the dolls we're looking at when we gaze at the night sky. Let's start with Earth. It's a rocky ball floating in space, one of nine planets circling our sun. It's about 8,000 miles across. Contrary to popular belief, people have known it was round for thousands of years—in fact, the ancient Greek Eratosthenes developed reasonably accurate ways of measuring Earth's circumference

EYE IN THE SKY

Glimmering like some Masonic vision, the Hourglass Nebula encircles a dying white dwarf, a type of star, just to the left of the "pupil" of its "eye." Scientists study such planetary nebulae to learn more about the ejection of stellar matter during the slow death of stars that are similar to our sun. A young nebula, the Hourglass lies about 8,000 light-years from Earth.

THE COSMOS: TWO VIEWS

Map of the entire sky as seen in infrared light (above) shows "fossil radiation," a background glow due to cosmic dust that has been warmed by all the stars that have ever existed. Many of these stars cannot be seen even through optical telescopes, possibly because intervening clouds of dust obscure them, or because "light pollution" from our own galaxy interferes.

The lower map shows the cosmos with all foreground light from our solar system and the Milky Way removed. Such maps promise to help scientists learn how stars and galaxies took shape and evolved in the wake of the Big Bang.

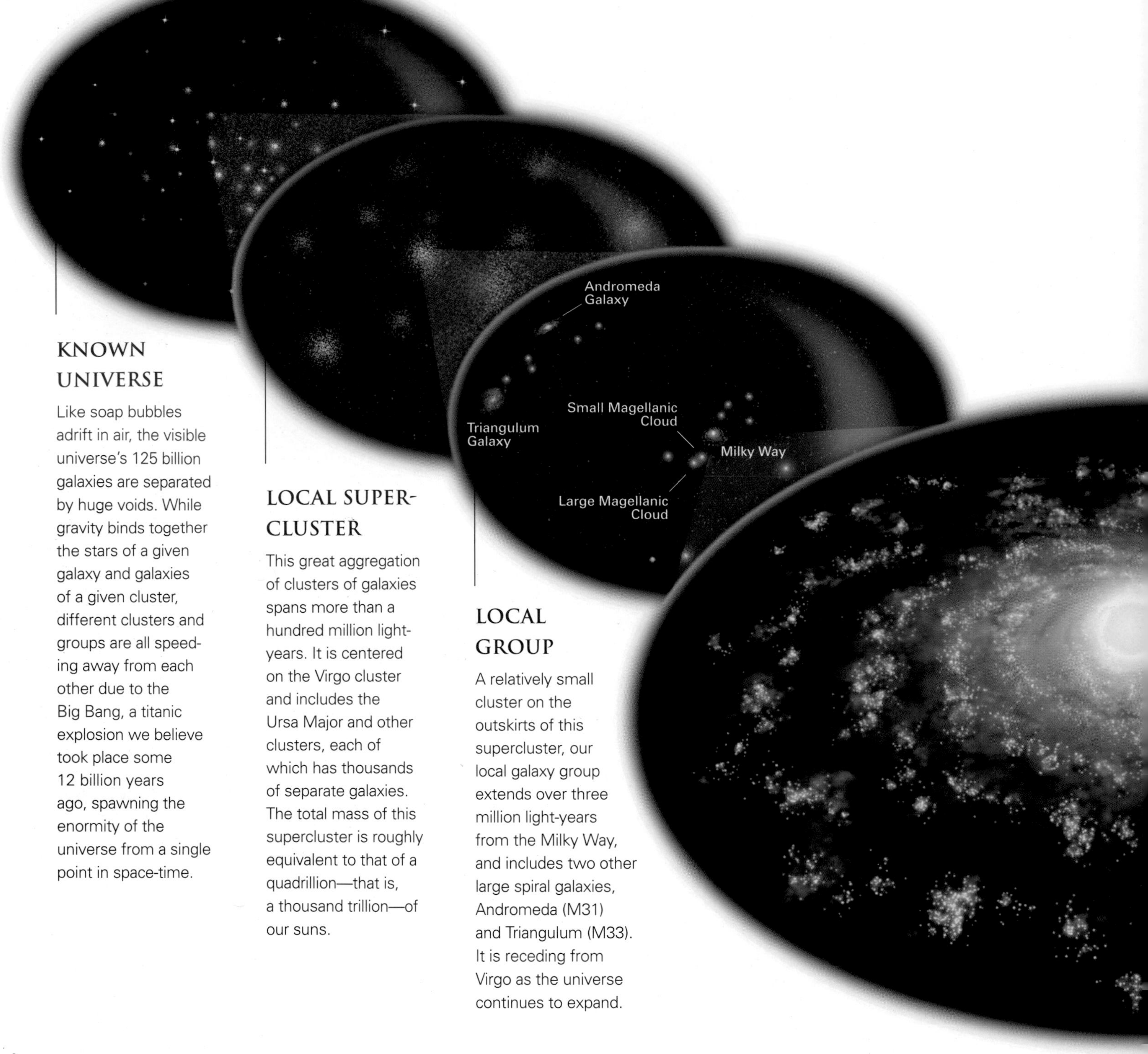

KNOWN UNIVERSE

Like soap bubbles adrift in air, the visible universe's 125 billion galaxies are separated by huge voids. While gravity binds together the stars of a given galaxy and galaxies of a given cluster, different clusters and groups are all speeding away from each other due to the Big Bang, a titanic explosion we believe took place some 12 billion years ago, spawning the enormity of the universe from a single point in space-time.

LOCAL SUPER-CLUSTER

This great aggregation of clusters of galaxies spans more than a hundred million light-years. It is centered on the Virgo cluster and includes the Ursa Major and other clusters, each of which has thousands of separate galaxies. The total mass of this supercluster is roughly equivalent to that of a quadrillion—that is, a thousand trillion—of our suns.

LOCAL GROUP

A relatively small cluster on the outskirts of this supercluster, our local galaxy group extends over three million light-years from the Milky Way, and includes two other large spiral galaxies, Andromeda (M31) and Triangulum (M33). It is receding from Virgo as the universe continues to expand.

Andromeda Galaxy

Triangulum Galaxy

Small Magellanic Cloud

Milky Way

Large Magellanic Cloud

THE BIG MATRYOSHKA

Truly colossal in scale, the cosmos we can see spans roughly 28 billion light-years, or about 165 sextillion miles (that's 165 billion trillion, or 165,000,000,000,000,000,000,000). How to fathom the unfathomable? By splitting it into smaller and smaller subunits that nest within each other, just like Russian *matryoshka* dolls. Some 125 billion galaxies occupy the visible universe; each of them, like our home galaxy, the Milky Way, is an "island universe" consisting of billions of stars. Galaxies occur in relatively small groups, in larger bands called clusters, and in still larger, filamentary aggregations known as superclusters. They contain enormous amounts of matter, yet are arrayed at the periphery of even more enormous voids of empty space.

MILKY WAY GALAXY

About 100,000 light-years in diameter, our home galaxy contains a few hundred billion stars, concentrated in a bulging core and flat, spiral arms. Just as planets circle the sun, the Milky Way's member stars orbit the galactic core, believed to enclose a gigantic black hole.

LOCAL STARS

Positioned in the Orion arm of the Milky Way, our local stellar neighborhood includes stars that exist within 20 to 30 light-years of Earth. Most are too dim to be seen without telescopes, but a few, such as Sirius, are familiar even to casual stargazers. The nearest star to our solar system, Proxima Centauri, is a red dwarf some 4.3 light-years away. It has only a tenth the mass of the sun, itself a very ordinary star in an ordinary spiral galaxy.

OUTER SOLAR SYSTEM

The five outermost planets orbiting our solar system include the four so-called gas giants—huge, cold, gaseous orbs with rings and numerous moons—and the rocky dwarf we call Pluto. Beyond the gas giants lie the flat Kuiper belt and the far larger, spherical Oort cloud, both of which harbor comets. We believe that all the planets, as well as the comets and the sun, condensed about 4.5 billion years ago from the disk-like solar nebula, a great cloud of dust and gas.

INNER SOLAR SYSTEM

Four rocky "terrestrial planets" orbit nearest the sun, accompanied by thousands of far smaller but equally rocky asteroids that occupy the asteroid belt between Jupiter and Mars. Earth, the only known abode of life, is outstanding also for its surface aggregations of liquid water as well as its volcanism, seafloor spreading, mountain building, continental drift, and other dynamic processes.

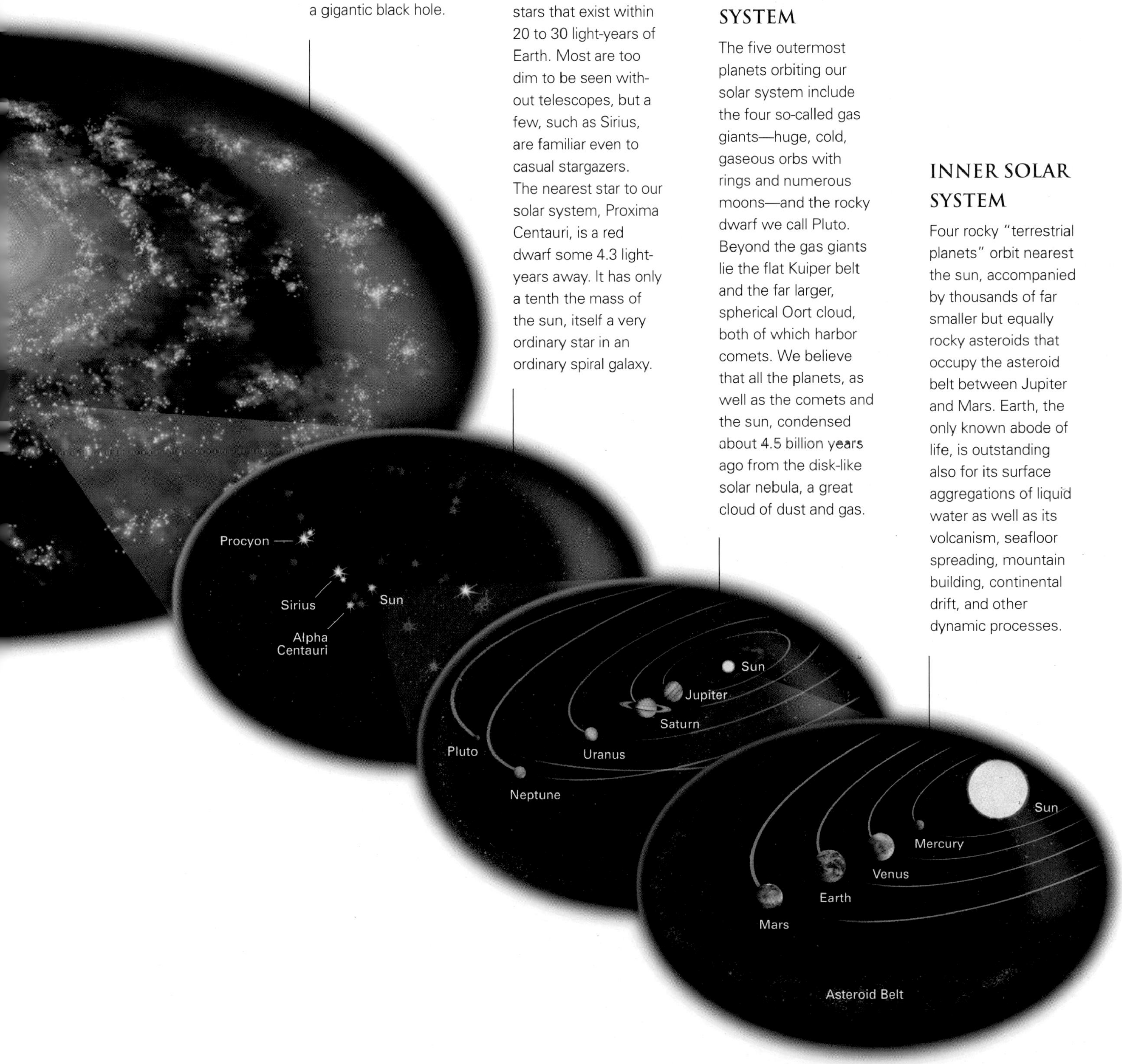

Procyon
Sirius
Sun
Alpha Centauri
Sun
Jupiter
Saturn
Uranus
Pluto
Neptune
Sun
Mercury
Venus
Earth
Mars
Asteroid Belt

Art by Don Foley (Milky Way illustration by Kenneth Eward/BioGrafx) by 250 B.C. Our home planet is unusual in many ways, but perhaps its most unusual attribute is that it is the only place in the universe where we are sure that life has developed.

We've all seen photographs of Earth floating in space, so it shouldn't be too hard to think about it as a world surrounded by emptiness. The hard part is to fathom just *how* empty that space is, even within the relatively small confines of our solar system. Let's say you're in a 747 jet going 600 miles per hour, and—for the sake of argument—that the jet can fly through space just as it flies through air. It would take over two weeks to reach the moon, our nearest neighbor. If you could stay in your seat that long, the plane would take over 17 years to get to the sun, and almost seven centuries to get to Pluto, the outermost planet. Given that the same 747 can deliver you to almost any point on Earth in a matter of hours, we begin to see what is involved in going from the smallest doll (Earth) to the next largest (the solar system).

Let's look at the solar system another way. Suppose we could shrink it down uniformly, so that the Earth was just the size of a grapefruit. Then the moon would be a walnut about twelve feet away; the sun would be a sphere as big as a four-story building located a mile distant. The other planets, swimming in the emptiness of space, would be hard to see. Pluto would be an all but invisible acorn almost 37 miles out. No wonder it's taken the human race millennia to come to terms with the enormity of our solar system.

But even these distances pale compared to the enormous ones that separate our solar system from other stars in our galaxy, the next largest doll. After all, for all of its emptiness, our scale-model solar system with its grapefruit-size Earth would fit comfortably within a major metropolitan area. Yet on the same scale, Proxima Centauri, the nearest star to us apart from our sun, would be over 250,000 miles away! (Its actual distance is 25 *trillion* miles, a span almost impossible to fathom.) If we want to comprehend the layer of the matryoshka that represents most of the stars we see, we're going to have to shrink the scale again.

Imagine, then, that we reduce not just Earth but the *entire solar system* to fit inside our standard grapefruit. Now the nearest star would be just over a half-mile away. But the Milky Way, our spiral home galaxy that includes billions of stars spread across 100,000 light-years of space, would span more than 12,000 miles—farther than from Tokyo to Buenos Aires.

But, of course, other galaxies exist beyond the Milky Way—each as magnificent and complex as our own. Together, they represent the final doll in our matryoshka, the ultimate outer reaches of our known universe. At this point, analogies start to fail. If we could reduce the 100,000-light-year span of the Milky Way itself down to the size of a grapefruit, then Andromeda, the nearest large galaxy to us, would be a slightly larger starry spiral turning in space a mere ten feet away. The Virgo cluster, a gigantic collection of galaxies that figures prominently in our exploration of the outer reaches of the universe, would be somewhat less than the length of a football field away, and the farthest galaxy that astronomers have yet found with their best telescopes would be about 12 miles distant.

Beyond that is still the unknown. We write here, as medieval mapmakers used to write in the blank spots on their maps, "Here be Lions." In time, perhaps we'll be able to make a few guesses about what might exist beyond our current reach, but for the moment we have to accept that our knowledge is not complete. Despite all the knowledge our forebears amassed in past centuries, there is more to learn, more to discover.

One of my favorite Greek thinkers is a rather obscure philosopher named Archytas of Tarentum, who lived around 400-350 B.C. and was a contemporary of Plato and a follower of Pythagoras. He tackled a difficult problem—the question of whether the universe was finite or infinite—in a very creative way. Suppose, he argued, that it was finite. Then it would have to have an edge. Imagine walking to that edge and extending an arm or a staff outward. There are, Archytas postulated, only two things that could happen. Either the staff reaches beyond the edge, or something deflects it and turns it back. Since, by definition, there can be nothing in the void beyond this finite universe, there can be nothing to deflect the staff. Wherever it reaches, you could walk to that point—in theory at least—pick up the staff, and reach out again. No matter what point you select for the edge of the universe, logic dictates that you can always find a place farther out. Therefore, Archytas determined, the universe must be infinite, without end or boundary.

This charming argument wouldn't pass muster today—Archytas didn't have any idea of what Albert Einstein would teach us about the curvature of space or the unity of space and time. Nevertheless, the story of Archytas provides us with a useful way to consider how our ideas about the universe in which we live have changed. Every once in a while a pioneer with a name like Nicolaus Copernicus or Edwin Hubble comes along with his staff and shows us that we live in a universe far bigger and more mysterious than anything we had previously imagined. We relearn the lesson that, to quote turn-of-the-century British biologist J.B.S. Haldane, "The universe is not only queerer than we suppose, but queerer than we *can* suppose."

IT'S OFTEN HARD FOR MODERN URBANITES TO UNDERSTAND THE FASCINATION THAT the night sky held for people before the invention of artificial lighting. Look at the heavens from any downtown today and you will probably see the moon and a few of the brightest stars, but they will seem pale and insignificant compared to the man-made lights around you. You have to go to the country, far from city lights, to see the sky as our ancestors saw it, blazing in splendor with stars so luminous they seem almost near enough to touch. Such displays used to be the birthright of every human being. Small wonder, then, that speculation about the universe and Earth's place in it has occupied humanity since the very dawn of history.

At first, people took the evidence of their senses at face value. What could be more obvious than the "fact" that Earth was flat and the sun, moon, and stars moved around it? But by the first millennium B.C., astronomers figured out that Earth wasn't flat but round. The evidence was varied—watch a ship move out from shore, and the hull always drops first below the horizon, followed by the sails. On a flat Earth, departing ships would just get smaller and smaller until they disappeared. Early astronomers also noticed that celestial events such as eclipses took place at different times of day in different places—what we now call different time zones. On a flat Earth, such events would take place at the same time everywhere. (I often tell my students the simple fact that they can sit in twilight on the East Coast and watch a football game being played in full sun in California is a modern version of this early proof of Earth's shape—if Earth were flat, the sun would set everywhere simultaneously.)

But even though educated people established Earth's shape long ago, the second common-sense notion about our planet—that it is the stationary center of the universe—persisted. The same astronomers who reasoned that Earth was round put together a picture of the universe in which the stars and planets circled Earth by means of a complex framework of crystal spheres. The system worked well enough to predict things such as eclipses and planetary positions, so that's more or less where things stood until the late Middle Ages.

In 1543, Polish cleric Nicolaus Copernicus published a book called *On the Revolutions of the Heavenly Spheres*. Written in dense prose and filled with complex calculations, this book was an unlikely candidate for the term "revolutionary"—yet that's just what it was. For its author produced a model of the universe every bit as complex and accurate as the one developed by the Greeks, but one in which the sun stood at the center of the solar system, while the Earth and other planets orbited it.

This Copernican universe had intellectual consequences far beyond what contemporaries of its inventor could imagine. If Earth was in fact moving, the stars should change their position with each passing season, just as telephone posts appear to move as you drive by them. Since stellar positions don't appear to change, then stars must be very far away—farther than anyone had ever dreamed. Suddenly, instead of living in a cozy, Earth-centered universe, humans found themselves adrift in a vast emptiness, the stars unimaginably distant.

When telescopes got good enough, a few centuries after Copernicus, astronomers were able to detect the apparent motion of the stars caused by the movement of the Earth in its orbit. In 1838, German astronomer Friedrich Bessel used this apparent motion to measure the distance to a particular star. His measurement established once and for all the scale of distances outside of our own solar system and banished forever any conception humans might have had that our own planet occupies a central position in the universe.

Less than a century later, American astronomer Edwin Hubble pushed the boundaries farther still. Working with what was then the state-of-the-art telescope on Mount Wilson in southern California, Hubble succeeded in measuring the distances to several neighboring galaxies. Not only did he establish that galaxies are separated by millions of light-years of empty space, he also found that distant galaxies are all moving away from us. This universal expansion, which now bears Hubble's name, forms the basis for another great insight— the concept that the universe began at a specific time and has been expanding ever since. This so-called Big Bang scenario is now one of the cornerstones of modern astronomy. In fact, working out the details of this expansion—whether it is slowing down or speeding up, for example—remains a frontier of scientific research today. On top of the *spatial* organization of the universe we've been discussing, then, Hubble suggested a *temporal* one—a time line that takes us from the earliest stages to the present, with astronomers debating the ultimate fate of the universe.

It was in this roiling, expanding, hierarchical universe that the human race evolved and eventually developed the science that now allows us to understand the night sky. But perhaps the most important thing we've learned is that we are not separate from this spectacle but are an integral part of it. Sometimes our connection to the universe is easy to detect. Subatomic particles from space continuously strike our planet and pass through our bodies. Meteorites and debris from comets and asteroids rain down, adding 40,000 tons to Earth's bulk each year. Large impacts occasionally create widespread havoc, such as the extinction of the dinosaurs.

Sometimes the connections are more subtle (though no less important). For example, the calcium in our bones and the iron in our blood were not created on Earth but forged in the explosions of now-distant stars, billions of years ago. All of us are, quite literally, at one with the stars. We are not just observers in our universe, but participants.

So with this background, let's board our starship and journey out into the magnificent universe we inhabit. Let's see for ourselves just what's out there and what it means to us. ●

TWO PIONEERS

At the eyepiece of the Palomar telescope in California, American astronomer Edwin Hubble (1889-1953) showed that other galaxies indeed exist, and that our universe is expanding. His namesake space telescope (opposite) was refurbished in December 1993 by astronauts high above Australia's west coast.

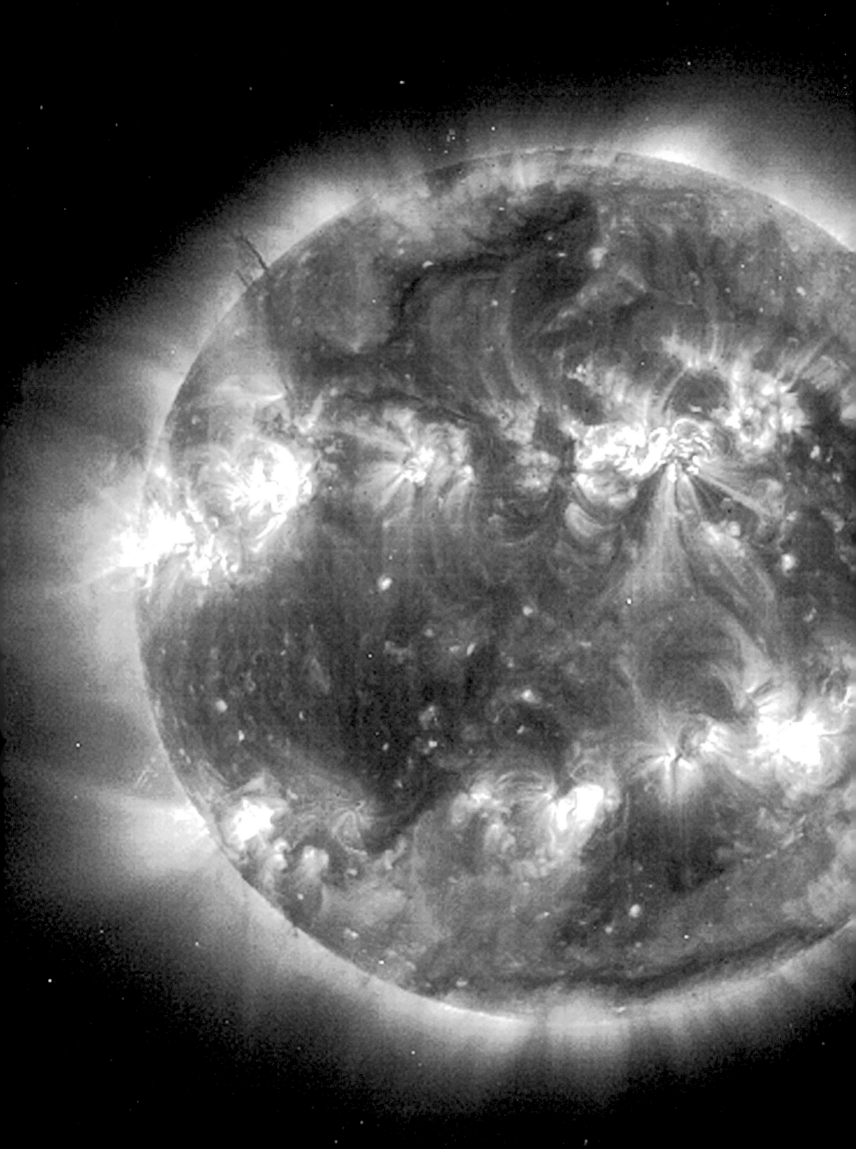

BIRTH OF THE SOLAR SYSTEM

THE SUN

Turbulent heart of the solar system, our home star took shape about 4.5 billion years ago, as interstellar gas and dust collapsed into itself and began to heat up. Similar scenes of star birth unfold throughout the universe today; a Hubble Space Telescope image (right) shows gigantic, funnel-shaped clouds interacting with young, hot stars in the Lagoon Nebula.

[PREVIOUS PAGES]
Viewed in the extreme ultraviolet range, the sun's billowy corona, or outer atmosphere, surrounds a tumultuous surface.

FIVE BILLION YEARS AGO THERE WAS NO SUN, NO EARTH, AND, of course, no human beings to be aware of their absence. All that existed in the area of the galaxy we now call home was an enormous, diffuse cloud of interstellar dust and gas. It consisted mostly of helium and hydrogen, the basic constituents of the universe even today, but here and there were bits of dust—tiny grains of heavier material that had been spewed out by dying stars during the long reaches of time. Had we been there to see it, the cloud wouldn't have looked spectacular—just a wispy, diaphanous collection of stuff rotating lazily in the enormity of space.

But the cloud wasn't uniform—far from it. It was lumpy, and had we been able to examine it more closely, we would have seen that one area of it had, by chance, a denser collection of material than others. Since the gravitational force exerted by an object depends on its mass, the gravitational force exerted by this relatively high-density area was correspondingly above average. Consequently, it tended to pull more material toward it, further increasing its mass. As a result, its gravitational force increased, more matter was pulled in, increasing the force even more, and so on. Eventually, the most massive region of the cloud began pulling everything toward it, and the entire cloud started to collapse.

That caused two things to happen. First, the amount of material in the center grew enormously, creating a huge and increasingly dense ball in space. As its density increased, so did its temperature. You can see the same effect in everyday examples like a bicycle pump: Inflate a tire by hand, and the pump gets warm. That's because you're compressing the air, making it denser and causing its temperature to increase, which in turn heats the pump.

The second thing that happened was that, like an ice skater that draws in her arms as she goes into a spin, the cloud's initially slow rotation sped up as it decreased in size and increased in density, pulling outlying material into a flat disk in the plane of rotation.

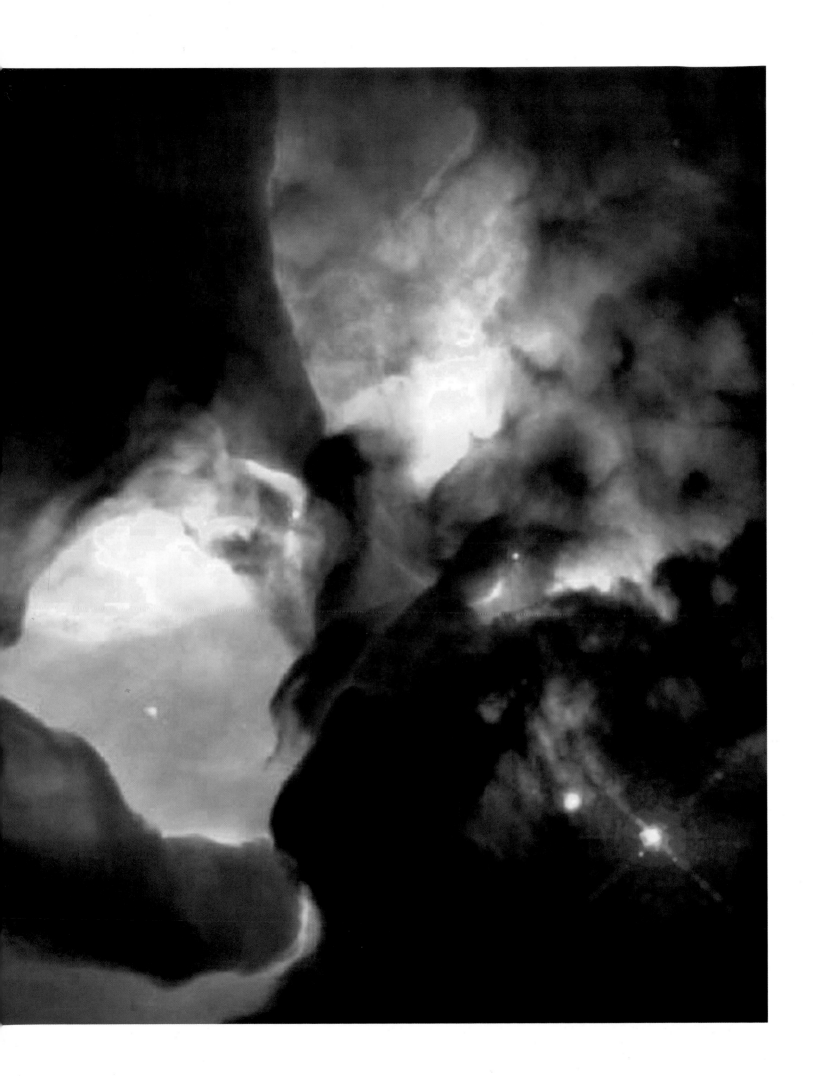

Eventually, the system began to look something like a pancake with a big scoop of butter at the center. Both parts consisted of pretty much the same stuff, except that the disk contained a lot less, spread over a much larger space—only about one percent of the material in the original cloud. Everything else was now in the center.

The future of these parts—the central body and the disk—would be very different. The central object became a star, our sun, while the material in the disk formed the planets and their moons, asteroids, and the comets that today make up the most striking parts of our solar system. Let's look at the sun first, then come back and talk about what happened in the disk.

AS THE CENTRAL SPHERE CONTRACTED AND MORE MATERIAL fell in, it got hotter and began to glow. Gravity inexorably kept pulling the ball in on itself, further compressing and heating the in-falling clouds of material around it. Eventually its temperature got so high that electrons were stripped from the atoms of this cosmic matter, leaving the basic material of our sun-to-be a seething mass of negatively charged electrons and positively charged atomic nuclei. Physicists call such a collection of highly charged particles a plasma, and usually designate it as a fourth state of matter, as important as the more familiar states of solid, liquid, and gas. Today, most of the matter in the sun and other stars remains in the plasma state.

As core temperatures in the whirling ball of gas continued to increase, the constituent particles moved faster and faster. By far the largest component of the nascent sun was hydrogen, whose atoms consist of a single proton bound to one electron. As we have seen, high temperatures quickly stripped off electrons, leaving the protons racing around and colliding with each other. At low temperatures, protons don't collide because each carries a positive electrical charge, and like charges repel each other. But once temperatures began to approach the sun's current central temperature of 15 million degrees Celsius, those protons were moving so fast that they overcame their mutual repulsion and began to interact with one another. This was the start of nuclear fusion, the true beginning of our sun.

The net effect of fusion in the sun's core is to take four protons and produce the nucleus of a helium atom—two protons and two neutrons—along with some other particles and a lot of energy. This energy heats up the core even more and sets up an outward pressure that counteracts the inward pull of gravity. The enormous temperatures and pressures needed to support fusion exist only in the sun's core. Surface material, while still largely in plasma form, isn't nearly hot enough to support these reactions.

After a brief period of intermittent flare-ups (much like a balky car on a cold morning), the sun finally became a star, consuming hydrogen in fusion reactions that produced helium and enough energy to hold off collapse due to gravity. These reactions also produced the light and heat that eventually made life possible on Earth.

Today, our sun consumes hydrogen at a rate that is astronomical in every sense. Each second, about 700 million tons of the stuff is converted into helium. The sun has been using up hydrogen at an increasing rate since it was born about four-and-a-half billion years ago, but it still has enough to continue for another six billion years or so. As long as the hydrogen supply in its core lasts, the sun will continue to be the safe, stolid, steady star we know it to be.

A FEW FACTS ABOUT THE SUN:

- It's over 800,000 miles across. A 747 jet traveling at 600 miles per hour would take more than six months to circle its equator once. You could line up more than 100 Earths, each touching the next, and still not span the diameter of the sun. In fact, there's so much matter in the sun that you could make more than 333,000 Earth-size planets out of it.
- The sun's surface temperature is about 5500°C. This is well above the melting point of metals like tungsten and the boiling point of materials like carbon. (Its central temperature, as we've seen, is far higher: 15 million degrees Celsius.)
- While the density of the sun's core is greater than that of any materials here on Earth, the sun's overall density is not even 50 percent higher than that of water. This means that if it weren't for considerations of temperature, you and a friend could easily load a bathtub-size block of average solar material into a pickup truck and drive it around town.

A given bit of energy generated by fusion in the core of the sun takes an incredibly long time to work its way to the surface: Estimates range from more than ten thousand to one million years. That's because the gamma rays created by the nuclear fires at first bounce from particle to particle inside the sun, gradually working their way outward much as a harried passenger might bounce through a crowd in a tightly packed airport. Eventually, about 70 percent of the way out, the sun's pressure drops to the point where solar material can "boil." Huge amounts of superheated gas rise toward the surface, radiate away their energy and cool, then drop back down to be warmed again and repeat the cycle, over and over.

Once our packet of solar energy gets to the sun's surface, it only takes eight minutes for its light to reach Earth. It's the voyage from the core to the surface that takes millennia. So, next time you warm your face in the first sunlight of spring, pause for a moment to consider that the heat you feel originated not instantaneously or

even just a few minutes earlier, but long before the first human beings started planting crops for food.

In any case, the light that shines out from the sun—the stuff that gives you your suntan during the summer—comes from a glowing surface shell called the photosphere, the yellow disk we see. The temperature of the photosphere is about 5500°C. Outside of this ball are layers of successively thinner and hotter atmospheres whose light is normally lost in the photosphere's glare, but can be seen during eclipses. The chromosphere, for example, is a layer about 600 miles thick whose hydrogen gas gives it a bright red color that becomes visible only when the main body of the sun is blocked out.

The corona, that wispy but stunning pearl-white glow that can be seen fringing the sun at the moment of total solar eclipse, stretches millions of miles into space. Its temperature is over a million degrees Celsius, and it merges into a stream of particles that flow out from the sun, past the planets, and into interstellar space. This is the so-called solar wind, a flow of charged particles that compresses Earth's magnetic field on the "upstream" side, drags it out on the "downstream" side, and causes geomagnetic storms that produce auroras and disrupt radio communications.

The existence of the solar wind, in fact, points out yet another important feature of the sun—the fact that it has a magnetic field. Much of what we can see at the solar surface depends on the interaction of that field with the sun's plasma. The key point is that plasmas and magnetic fields tend to get locked into each other, much the way grass gets frozen into the ice at the edges of a pond. If either the sun's plasma or its magnetic field moves, it drags the other along. In fact, the solar wind is itself a weak plasma, which is why it both affects and is affected by the Earth's magnetic field.

Seen up close, the sun's surface is anything but smooth—it looks more like an ocean on a stormy day. Here and there prominences rear up, gigantic columns and arches of material that can extend 100,000 miles above the solar surface. Prominences are associated with disturbances in the sun's magnetic field, which are caused by the turbulent churning of gas in the sun's outer layer and by the fact that, unlike the Earth, different parts of the sun rotate at different rates.

Imagine what would happen if Earth's crust at the latitude of Florida moved faster than it did at Quebec: Eventually the surface of the planet would stretch and distort. Earth doesn't do this, but the sun does. Constant twisting of the solar surface stretches and pulls on the sun's magnetic field until, like a rubber band that's just had too much, it snaps. Localized snappings throw out huge amounts of plasma, which follow the magnetic field. The result is a dramatic explosion of energy we call a solar flare. A large flare can release the energy equivalent of a billion one-megaton bombs.

KAMIKAZE COMET

On its way to oblivion, the comet SOHO 6— one of a class known as sungrazers—arcs toward the solar corona, where intense solar heat will destroy it. Both the comet and the corona are visible only because the overwhelming glare of the sun's central disk has been blocked out.

Astronomers believe that the more than 75 known sungrazers originated from a single larger comet that passed near the sun in the 12th century and experienced successive breakups.

In a spectacular solar event known as a coronal mass ejection, or CME, perhaps a billion tons of the sun's atmosphere are expelled into space at a million miles an hour. Yet another phenomenon associated with the solar magnetic field concerns sunspots—relatively dark areas on the sun's surface. (Please remember to never look directly at the sun; doing so can cause serious eye damage and even blindness.)

Sunspots actually are quite bright, for their temperatures range above 3500°C, which means they radiate considerable amounts of energy in their own right. They only appear to be dark because we see them against the far brighter disk of the sun. Sunspots were known to the ancients, but were thought to be the shadows of clouds. By 1610, Galileo was able to observe sunspots moving across the solar disk, and he correctly concluded that they were on the sun's surface. In 1843 an amateur astronomer in Germany, Heinrich Schwabe, pointed out that sunspots come and go in cycles. Today, we know those cycles last an average of 11 years, and they are associated with the normal distortions of the sun's magnetic field. As distortions progress, more and more sunspots appear until the field breaks up, reforms, and the sunspot cycle starts anew.

Although the appearance of sunspots has been tied to changes in the sun's magnetic field, we do not really understand why that field reverses every 11 years. As we shall see, our state of knowledge of how magnetic fields at the planetary or stellar scale are created is not very advanced. This is just one of many things we don't know about the universe we inhabit. Astronomers have wondered for centuries whether sunspots affect Earth's weather. Many have postulated such a connection. In fact, some 19th-century astronomers tried— unsuccessfully—to use their knowledge of sunspots to speculate on grain futures, believing that high sunspot numbers corresponded to poor climate. Today, claims for a sunspot/weather connection are considerably more modest, but it remains an open question as to whether this solar phenomenon affects everyday life on Earth.

TODAY, OUR OBSERVATIONS OF THE SUN AREN'T CONFINED TO Earth-bound telescopes but are made from orbiting satellites and interplanetary probes. In 1990, for example, the spacecraft Ulysses was launched, becoming the first human artifact to leave the plane of the solar system and make a circuit around the poles of the sun. It last orbited the poles in 1994-95, when the sun was at the lowest point in its sunspot cycle. When it comes round again, in 2000-2001, the cycle should be at its peak. Perhaps this craft, wandering uncharted areas like the Greek hero for which it was named, will help us unlock the riddle of the sunspot cycle.

In 1995, when Ulysses was completing its first transit of the sun's poles, we acquired another important new tool with which to study our home star: the Solar and Heliospheric Observatory, or SOHO. Unlike Ulysses, SOHO stays relatively close to Earth, orbiting a mere million miles away. It was launched and sent to the L_1 point— that is, the first Lagrangian point, a spot in space where the sun's and Earth's gravitational forces just offset each other. SOHO traces an orbit around that point in space, neither approaching nor retreating from Earth, its dozen or so instruments keeping an unblinking eye on our sun. SOHO has detected and measured subtle seismic waves across the churning, heaving solar surface. Earth-based observations with the Global Oscillation Network Group (GONG) allow astronomers to probe the sun's interior with accuracy. Just as we learn about the interior of our own planet by remotely monitoring seismic waves created by earthquakes, so too are solar scientists learning more about the sun's interior by studying the output of these two projects.

And what a world it is! Beneath the sun's seemingly placid exterior, massive flows of gas rise and fall faster than a jet can fly, carrying hot material to the surface and returning cooled gases to the interior to be reheated. Meanwhile, in the corona, gigantic masses of gases—big enough to easily envelop the Earth—spew into space in a steady outward flow, occasionally punctuated by great blobs of ejected matter. No doubt about it—there's more to the sun than you might think when you bask under its rays on a warm summer day.

LET'S GO BACK, FOR A MOMENT, TO THAT INTERSTELLAR CLOUD that gave rise to our complex, energy-producing sun. At the same time it was collapsing and heating up, the disk of debris surrounding it also began to evolve. Close to the center, the dust and gas of the disk was denser, so it heated up. Astronomers believe this occurred throughout the region presently occupied by Mercury, Venus, Earth, Mars, and the asteroid belt. Farther out, temperatures were much lower and the kinds of planets that formed there were very different.

In the inner region of the disk, light gases such as hydrogen and helium—as well as vapors from volatile substances such as water— were driven off. What was left behind—the stuff that wouldn't melt or boil away—was in the form of solid grains of various minerals. These bits began to collide and stick together, accumulating into boulders from about a mile to a few hundred miles across, which astronomers call planetesimals.

As these objects swept in orbit around the sun, they occasionally collided with each other. Some collisions reduced planetesimals to rubble, but most caused them to stick together. Current theories suggest that the inner part of the early solar system may have acted like a gigantic game of billiards played with sticky balls. The planetesimals grew larger over time; dozens of orbiting moon-size

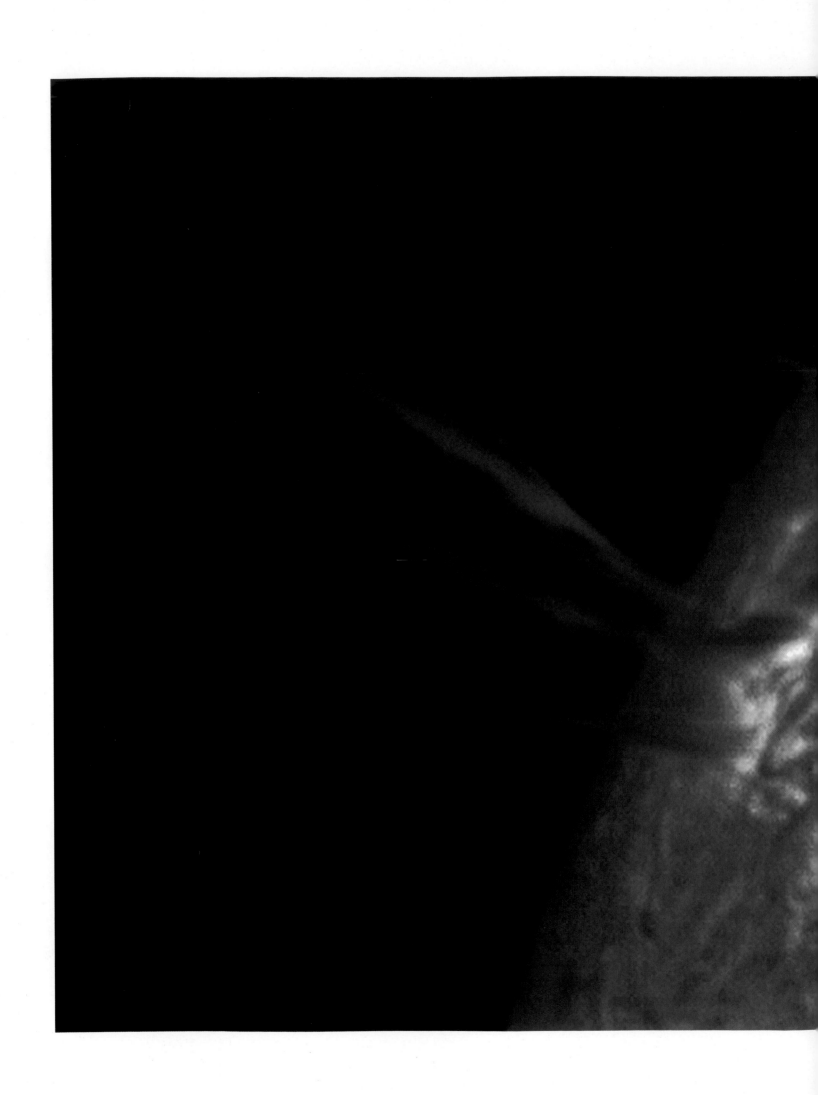

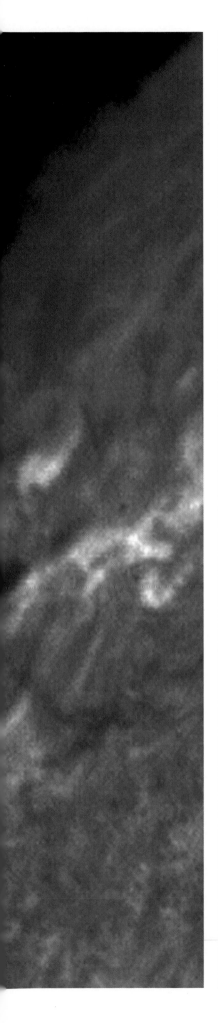

objects collided over and over, eventually becoming ever larger planets-to-be. Each increase in mass and gravitational attraction helped them sweep up remaining debris. The more matter that rained in on them, the more they heated up, at times melting all the way through. Current thinking holds that not only Earth but also all the inner planets of our solar system formed much this way, accreting from collisions, then heating up and melting, either totally or partially.

This early period, which extended roughly from 4.6 to 4.0 billion years ago, must have been pretty spectacular from the perspective of any of the newly forming planets. Giant chunks of debris would sweep in from space and smash into them, adding not only energy and mass but also changing their surfaces through the formation of craters. Astronomers often refer to this period as the great bombardment. On planets like the Earth, which developed an atmosphere and weather, these early craters quickly eroded away, but on bodies such as the moon, the absence of atmosphere and weathering allowed the craters to remain, mute testimony to a violent past.

Today, we believe that all planets nearest the sun—the so-called terrestrial planets, Mercury, Venus, Earth, and Mars—formed from the accumulation of planetesimals born of the "pancake," that is, the original protoplanetary disk. The fact that they experienced the same processes, however, does not mean they all wound up alike. For although they grew mainly from the accretion of rock and metals, the disk was probably far from uniform, resembling a tossed salad more than a puree. Planets that formed in different regions wound up with different sizes and different compositions, because different amounts and types of material existed at different distances from the sun. In addition, the way each planet evolved after its initial formation also differed, especially concerning such things as atmospheres and oceans. We'll discuss these differences in more detail in the next section, but they illustrate an important point about our universe: Processes governed by the same physical laws need not—and often don't—have the same outcomes. In fact, each of the worlds in our solar system is the product of a unique history, presenting us with a different set of puzzles as we try to understand it.

As planetesimals were colliding and merging in the part of the solar system nearest the sun, other events were taking place farther out, in the frigid region beyond the current asteroid belt. Here, the temperatures were so low that ices and even gases accreted with rocky material. Jupiter and Saturn grew so large that they attracted huge amounts of gas. So it is that the planets of our solar system fall into two classes—the hard, Earth-like ones near the sun, and the so-called gas giants farther out. They differ because conditions in the regions where they formed were so different. With this in mind, we'll next set out on a tour of our home system. ●

SOLAR ERUPTION

Gigantic spikes erupt from the sun's surface as extreme heat causes solar flares to release x-rays, ultraviolet light, and filaments of solar material. Such displays are often at the heart of even larger events known as coronal mass ejections or CMEs, in which billions of tons of superheated plasma race outward at millions of miles per hour.

THE SUN

Eerie majesty of a total solar eclipse darkens the daytime sky on
July 11, 1991, blocking the glare of the sun's surface and allowing the
delicate light of its corona, or outer atmosphere, to be seen. Primary
energy source for the solar system, the sun generates immense amounts
of energy by thermonuclear fusion, transforming hydrogen into helium.
Though obviously very special to us, our sun is average—in size, mass,
composition, age, and energy output—compared to other stars.

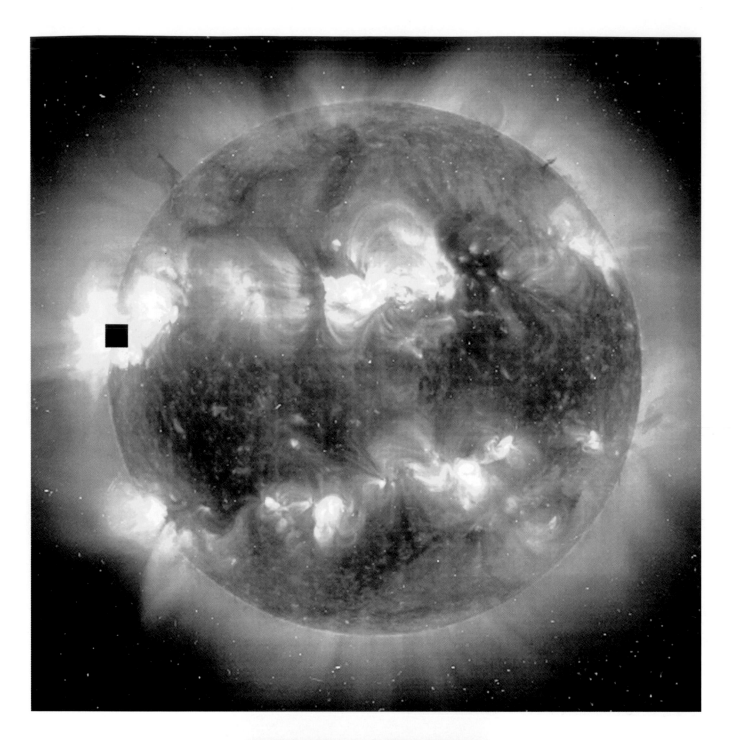

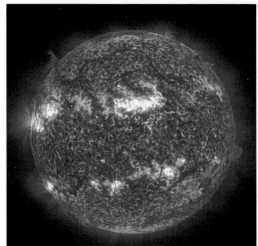

36

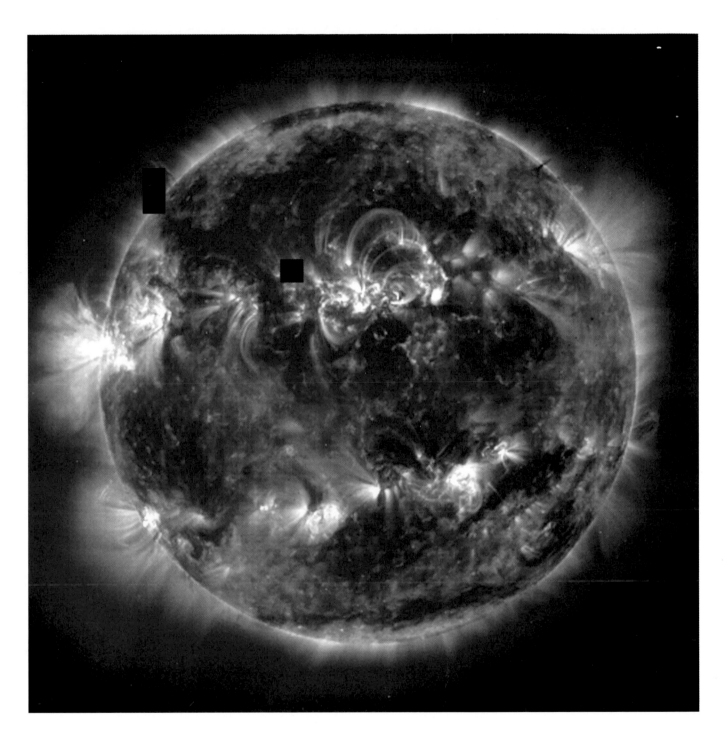

TRIO OF SUNS

Using different wavelengths of light in the extreme ultraviolet range and keying the data to visible colors, the Solar and Heliospheric Observatory spacecraft (SOHO) produced these three views. All highlight enormous convection patterns as the relatively cool solar surface interacts both with interior areas and with the diffuse but far hotter corona that fringes it. Data for the three were collected in November 1998.

SOHO operates about a million miles from Earth, focusing on the sun as it orbits the L_1 neutral gravity point—a spot in space where the gravitational forces of the sun and Earth perfectly offset each other.

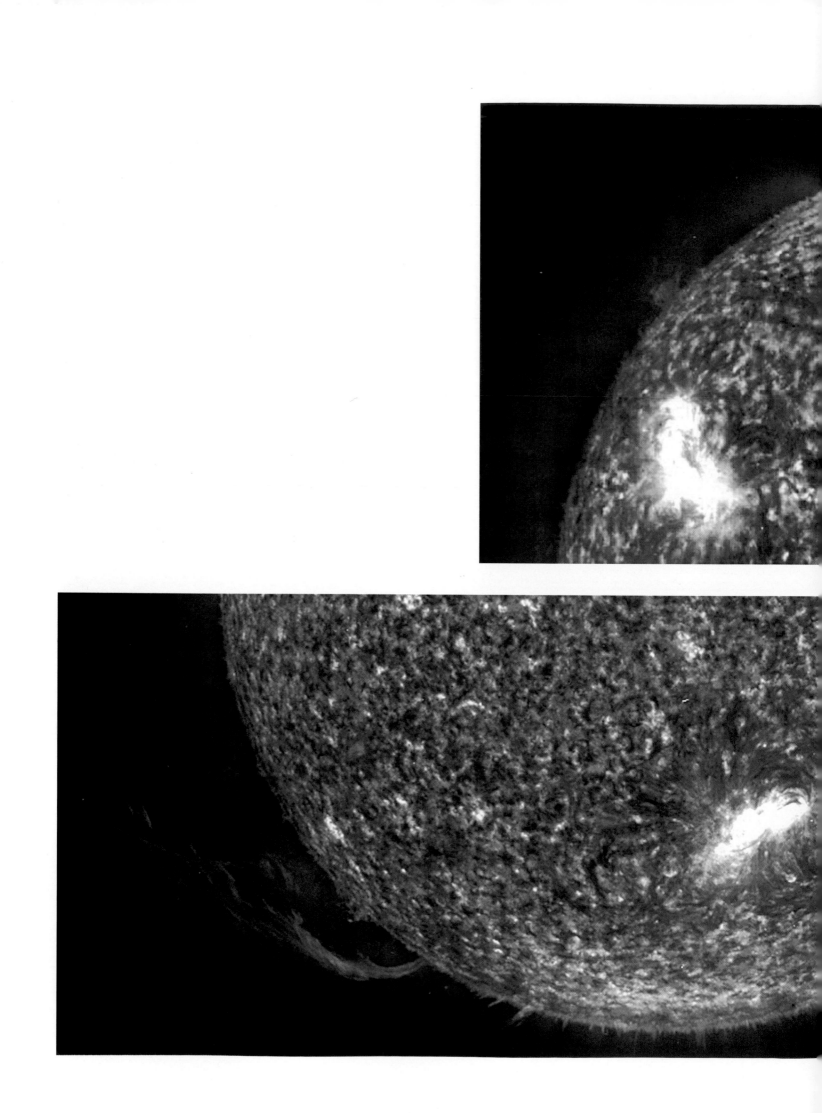

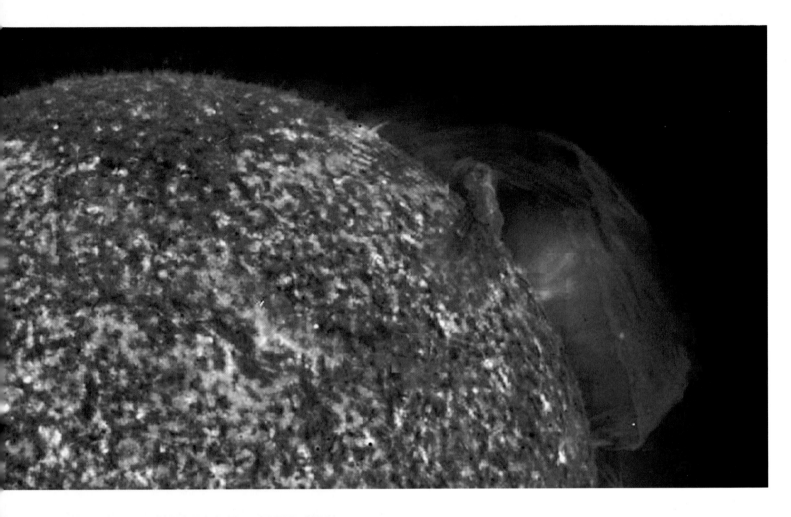

SOLAR PROMINENCES

Ever in turmoil, the sun's surface seethes with massive contorted arches known as solar prominences, which can extend 100,000 miles high. One especially large arch (above) bridges more than 200,000 miles of solar surface as it erupts, enough to cover 27 Earths. Twisting magnetic fields—caused by the convection of turbulent gases and differential rotation beneath the visible surface—suspend these clouds of relatively cool gas in the much hotter corona, sometimes expelling them at great speed (left).

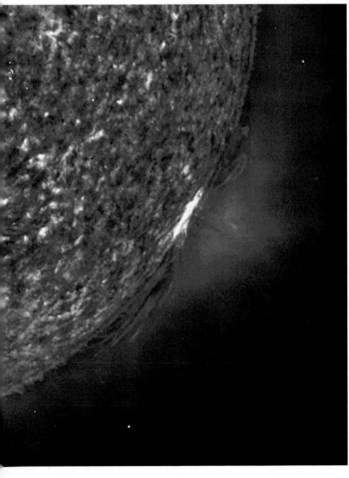

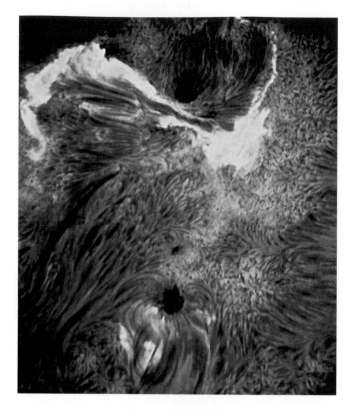

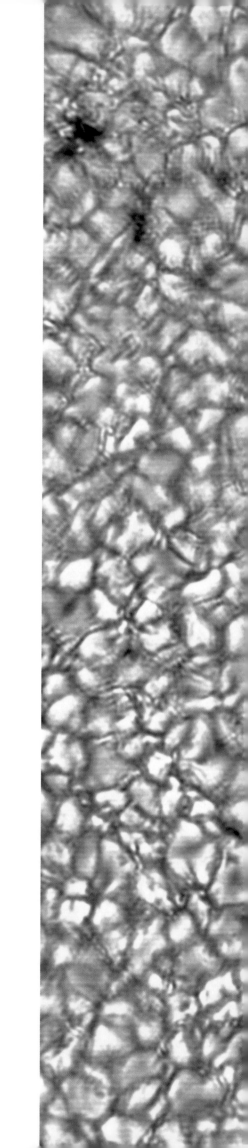

FOUNTAINS OF ENERGY

In addition to solar prominences, many other energetic phenomena mark the sun's seemingly smooth face. Solar flares (above) violently release tremendous amounts of energy and plasma into space within seconds.

Often occurring in pairs, sunspots (right), possess magnetic fields strong enough to inhibit normal convection. They appear dark to us because they are somewhat cooler than their granular surroundings— about 3500°C rather than 5500°C. Individual solar granules shown here, consisting of hot gas that has risen from below, are hundreds of miles wide.

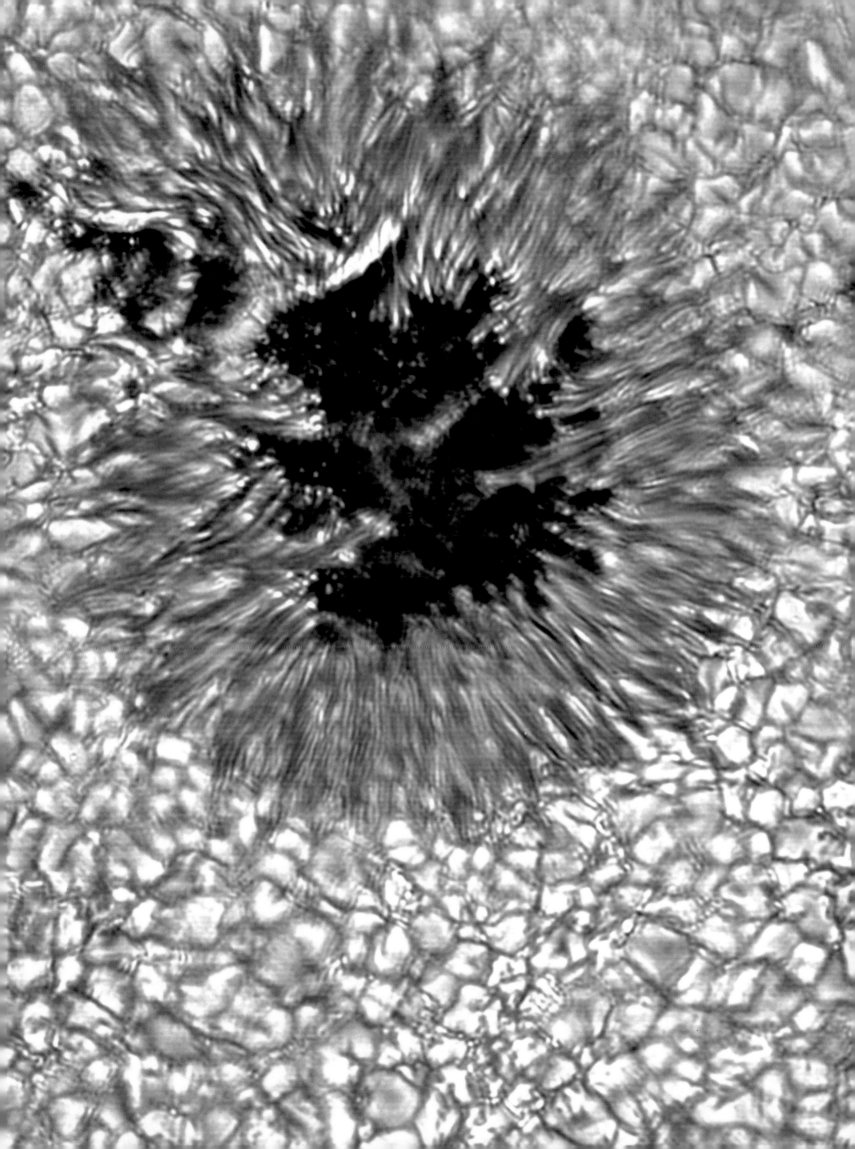

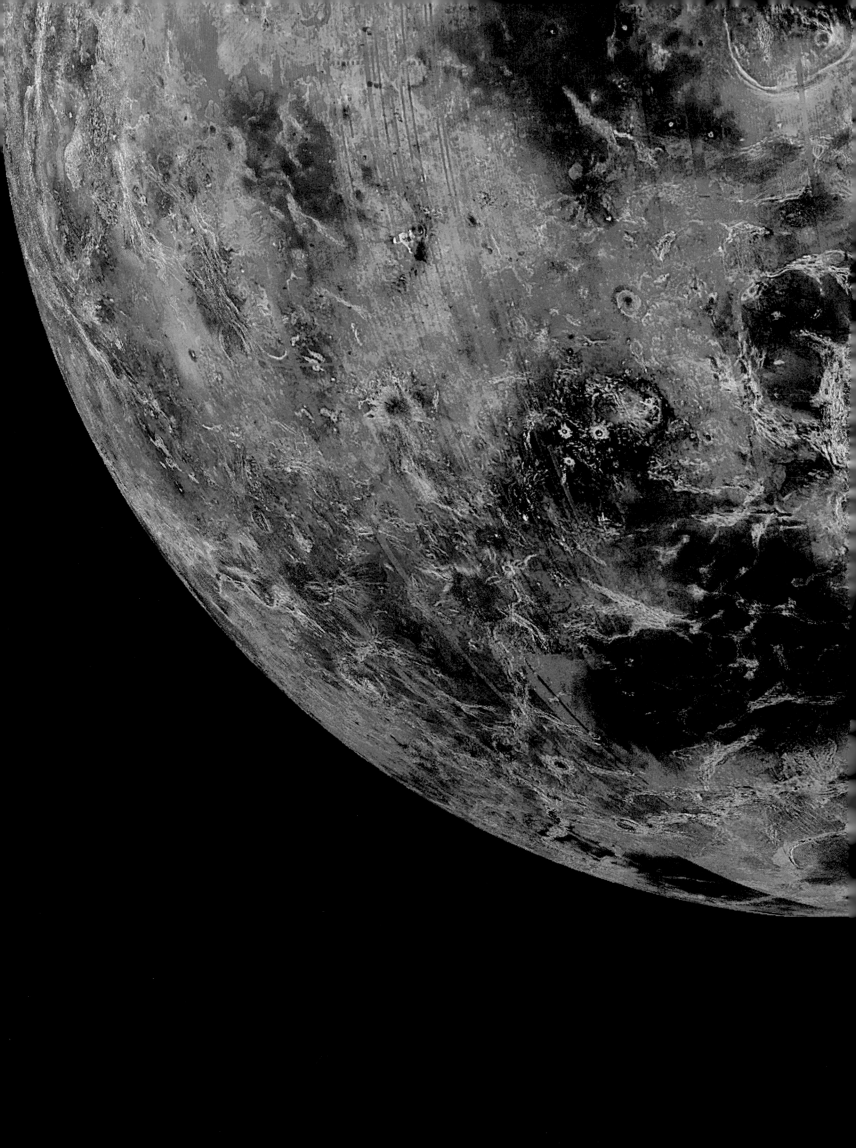

THE
INNER PLANETS

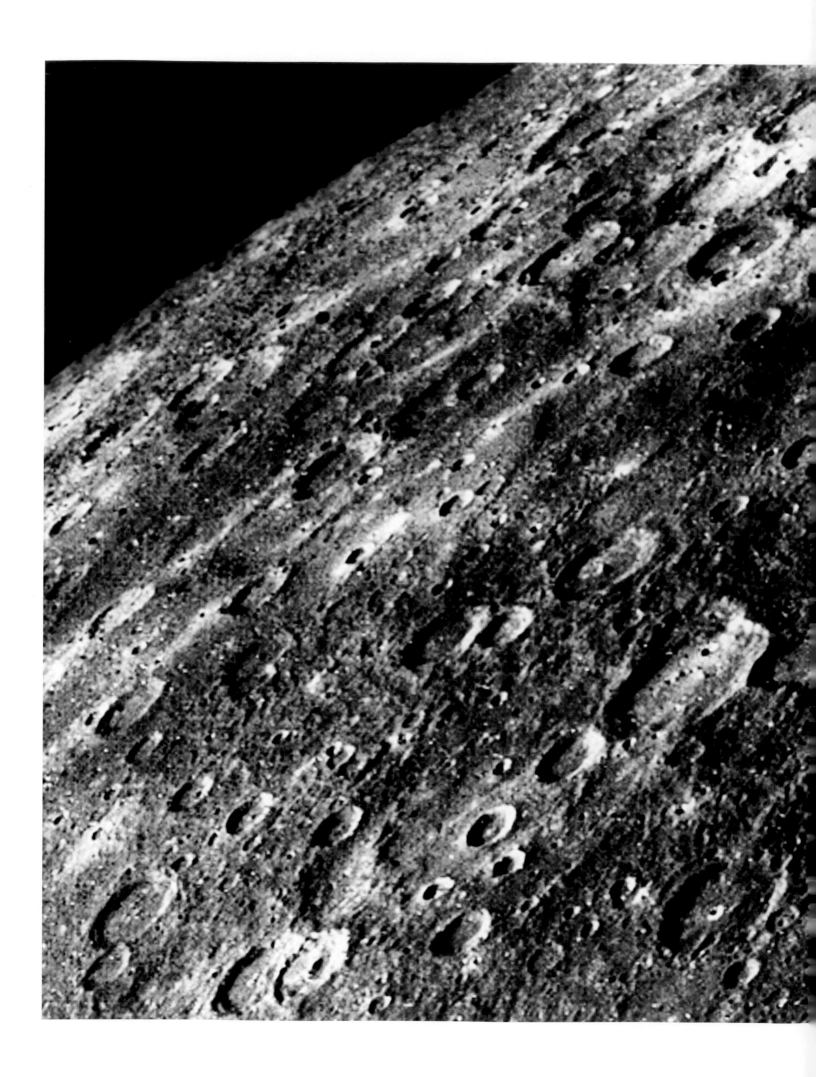

AS WE HEAD OUT FROM THE SUN, THE FIRST OBJECT WE encounter is Mercury, smallest of the terrestrial planets. Visible as a morning or evening star from Earth, Mercury makes a circuit of the sun every 88 days. Its swift motion through the sky encouraged the ancients to name it for the fleet-footed messenger of the gods.

It's hard to study Mercury because, from our perspective on Earth, it never gets more than 28 degrees away from the sun—about the angle formed by the hands of a watch at 11 o'clock—which means it tends to get lost in the solar glare. In fact, operators of the Hubble Space Telescope never point its instruments at Mercury for fear of accidentally damaging the sensitive equipment. So it is that Mercury, the planet nearest the sun, and Pluto, the smallest and most distant, share the distinction of being the least studied planets in our solar system.

So far, only one spacecraft has visited Mercury: Mariner 10, which made three flybys back in 1974 and 1975 and mapped about a third of the planet's surface. Mariner's photographs revealed a harsh world of rocky landscapes still dotted by thousands of craters made during the great bombardment that took place over four billion years ago. Scientists decided to name the planet's craters in honor of great artists, writers, composers, and musicians, including Mozart, Beethoven, Shakespeare, and others. Temperatures on Mercury range from about 450°C (hotter than the melting point of lead) on its sunlit side to minus 170°C on its dark side. That is the biggest temperature swing on any planet, and if Mercury had an atmosphere, you can bet that it would also have ferocious winds. But except for traces of sodium and potassium that percolate up from the planet's crust and bits of helium from the crust or from the solar wind, Mercury has no atmosphere—it's just too small a body to keep gases from escaping into space. And so it sits and bakes in the sun without a single breeze to break the monotony. Think of this planet as a hotter, rougher version of the Earth's moon.

MERCURY

Lacking almost any atmosphere, the sun's innermost planet endures 600-degree swings in temperature. Its heavily cratered surface (left) resembles the moon's, but its relatively high density indicates that its interior may be more like Earth's.

[PREVIOUS PAGES]

A decade of data contributed to this lavish false-color image of Venus, stripped bare of its cloud cover by radar, with each hue reflecting a different range of altitudes.

Scientists once thought that the time Mercury takes to go around the sun—that is, its year—and the time it takes it to turn on its axis—its day—were the same. Some great science-fiction detective stories pivoted on the "fact" that this planet always kept the same side turned toward the sun, much as our moon always faces Earth. In actuality, however, the length of one Mercury day is a little over 58 Earth days—exactly two-thirds of Mercury's 88-day year—so every spot on this planet gets some time in the sun. If you were to stand on Mercury, the apparent length of its day would seem longer still: Because this planet orbits the sun so swiftly while it turns relatively slowly on its axis, the interval between situations in which someone standing on a given spot on its surface would see the sun directly overhead is actually 176 Earth days—twice the length of Mercury's year!

Mercury's most impressive surface feature by far is a gigantic crater in its northern hemisphere. Called the Caloris Basin because it's in a place that gets a lot of heat and light from the sun, this structure is over 800 miles across—greater than the distance between New York and Chicago. Astronomers believe it is the remains of a giant impact that occurred about 3.8 billion years ago, an event so cataclysmic that the basin has a double ring of mountains around it, with the outermost being over 900 miles across. The floor of the basin is smooth, having been filled in by post-impact lava flows.

On Mercury's other side, directly opposite Caloris, there is a region of linear depressions and jumbled hills, some over a mile high. Scientists believe that when the Caloris impact occurred, gigantic shock waves ripped through the center of the planet, tearing up its surface and creating this still-chaotic terrain in the process.

But we don't know Mercury's life history in any detail. We believe it formed like the other terrestrial planets, from the accumulation of planetesimals, and was heated to the melting point. Heavy materials such as iron sank to the center where they formed a core, while lighter materials floated to the surface. This process, called differentiation, occurred in all terrestrial planets, and we'll see it again on our tour. After a relatively brief period as a ball of molten material orbiting near the sun, Mercury cooled and its outer surface solidified. Meteorites continued to impact it occasionally, releasing lava flows whose solidified remains we still see today.

Being a small planet, Mercury was—and is—ill suited to hold its heat against the cold of space. For many years, scientists assumed that it completely froze following its molten stage, locking even its interior into solid rock. It was a dead world, changed only by the occasional impact of a meteorite or comet. There are, however, two problems with this view. By measuring Mercury's gravitational effect on Mariner 10, scientists were able to get a very accurate measure of the planet's mass. Mercury turned out to be very heavy for its size, indicating that it contains a lot more iron, proportionately, than does the Earth. Why this is so remains a mystery.

In addition, Mercury's unexpectedly powerful magnetic field probably indicates that some iron in its core is still molten—as it is in the Earth's core. Again, there is no widely accepted explanation for this fact. And while various theories explain the presence of iron on Mercury and the strength of this planet's magnetic field, they remain speculation until we get more data.

One more thing: Mercury's high density cannot be explained by condensation and accretion alone. This planet may once have had more rocky material, which was vaporized and driven off by the sun. Alternatively, a massive impact could have removed much of its lighter exterior. I'm afraid that we will have to wait for another space probe to learn more about this tiny, enigmatic little brother to the Earth.

ASIDE FROM THE MOON, CLOUD-WRAPPED VENUS IS THE brightest object in the night sky. It is often visible as an evening or morning star, even when you're looking up through the murk and lights of a city. Venus is bright for two reasons—it is relatively close to the Earth, and it is surrounded by a thick layer of white, highly reflective clouds. Since planets shine by reflected light, the clouds of Venus add considerably to its luster. Its brightness and inviting, mysterious appearance in the evening sky caused the ancients to name this planet for the goddess of love.

In many ways, Venus is the most Earth-like of the terrestrial planets. It is 82 percent as massive as our planet and has a radius only 5 percent smaller. Indeed, scientists used to consider Venus a warm and wet world, something like Earth's tropical wetlands. The "swamps of Venus" became a science-fiction staple throughout the mid-20th century. Today we know better. Those fleecy white Venusian clouds, so beautiful to look at, are actually a seething hell composed of droplets of sulfuric acid—the same as the stuff in your car battery. Temperatures at the surface, far from being comfortably tropical, are actually a torrid 500°C—hot enough to boil off any swamps that might ever have existed.

One more tidbit: Early measurements by radio telescopes on Earth established the fact that Venus rotates "backward" as it orbits the sun once every 243 days. That is, if you could stand on its surface and see through its dense clouds, the sun would appear to rise in the west and set in the east—the exact opposite of what happens on the Earth. This anomalous rotation is thought to be an historical accident, one that reveals little about the planet's fundamental nature.

Actually, we know a good deal about Venus despite its cloak of clouds, for it is one of only three heavenly bodies on which humans have landed spacecraft. (The moon and Mars are the others.) In 1970, the then-Soviet Union landed the first such vehicle on Venus. The

VENUS

Beautiful only at a distance, the solar system's hottest planet wears a shroud of dense clouds rich in sulfuric acid. Its atmosphere retains heat and creates a surface pressure 90 times greater than Earth's. Although similar to our planet in size and position, Venus has had a very different fate.

VENUS REVEALED

Landing on the planet in 1982, the Soviet Venera 14 spacecraft sent back bleak images of the surface: rocky scapes devoid of soil or even small pebbles that would indicate erosion, as occurs on Earth. The atmosphere beneath the Venusian cloud cover—mainly carbon dioxide—was transparent, allowing broad views clear to the horizon. Venera's sawtooth ring, visible here, helped stabilize its descent and absorb the shock of touchdown.

strategy for dealing with the planet's high temperatures was simple: Use the craft's long space journey to chill it, then gently drop it to the surface and have it radio back information until the heat finishes it off. In a series of landings involving four different spacecraft, the project yielded photographs of different parts of the Venusian surface. Some photos revealed flat plains studded with rocks and lava flows—rather humdrum compared to the fictional swamps. They also showed that the clouds don't extend down to the surface of the planet but stop about 30 miles up, leaving a clear layer of orange-colored sky. That color arises from the fact that, to reach the surface, sunlight must pass through a lot of atmosphere, which scatters the shorter wavelengths associated with blue and green light, allowing reds and oranges to dominate. The same thing happens at sunset and sunrise on Earth, when slanting rays from the sun traverse more air than they do at noon, thus their light is more scattered and red.

Following the early Soviet Venera probes, NASA's Pioneer (between 1978 and 1981) and Magellan (starting in 1990) spacecraft orbited Venus expressly to map its surface. Magellan, in particular, sent back dramatic high-resolution images of the surface beneath the clouds. We sometimes get blasé about what's involved in mapping a world millions of miles away. But I still remember the thrill when I visited NASA headquarters during the early days of Magellan and watched those images slowly coming back from another planet, line by line, gradually revealing Venus's long-hidden topography.

Somewhat less than two-thirds of the Venusian surface consists of relatively flat, rolling plains. The most dramatic feature is a continent-like mass in the northern hemisphere called Ishtar Terra. (In keeping with the ancient custom of identifying this planet with a feminine

deity, major geographical features on Venus are named after prominent women or female mythological figures; Ishtar was the Mesopotamian goddess of love and war.) Ishtar Terra is a high plateau, similar in terrain to Tibet but about as large as the United States. It is bounded by mountains, some of which stand over a mile taller than our own Mount Everest. What appears to be missing are mountain chains like the Rockies or Andes, geological boundaries like the San Andreas Fault and the Mid-Atlantic Ridge, and any evidence that molten magma from the planet's interior is rising to the surface.

How did a planet with so many overall similarities to Earth develop into something so different? Venus's "backward" rotation is easy to explain in a solar system where large bodies are assembled from smaller planetesimals. In such a process, each planet's rotation is determined by the angles and speeds with which the last few major planetesimals come together. All that is necessary to reverse the direction of rotation is that at some time in the final stages of assembly, a large and fast-moving planetesimal collided with Venus in just the right way. The same thing could happen on Earth today if, for example, a big and speedy asteroid coming from the east were to smack into eastern North America with enough momentum to cause the globe to reverse its motion. Although scientists tend not to like such ad-hoc explanations, when it comes to Venus's direction of rotation, it's the best they can do at the moment.

This planet's high temperature is also easily explained in terms of an increasingly familiar phenomenon on Earth—the greenhouse effect. The current Venusian atmosphere is about 95 percent carbon dioxide, which absorbs infrared radiation emitted by Venus and sends it back to the planet itself. Think of this dense atmosphere as a kind of gigantic blanket. Venus and Earth have about the same amount of carbon dioxide in their outer surfaces, but on Earth most of this gas has been taken into the oceans and, from there, into carbonate rocks and the skeletons of living things. This means that on Earth, carbon dioxide eventually winds up being stored either as a rocky mineral or as a dissolved constituent of the oceans. Venus is so much closer to the sun that its oceans (if indeed they formed at all) quickly would have evaporated. Consequently, all its carbon dioxide stayed in its atmosphere, where it contributed to planetary warming.

Thus Venus exemplifies a runaway greenhouse effect, in which atmospheric carbon dioxide raises temperatures to the point where any water present boils away, preventing the chemical reactions that might remove carbon dioxide from the atmosphere and lower the temperature. (Current debate over the greenhouse effect on Earth, by the way, centers on whether our home planet's temperature will rise a few degrees over the next century, not about whether it will turn into a searing hothouse like Venus.)

EARTH

Most favored planet, the third rock from the sun gleams blue with liquid water—necessary for life as we know it—and white with clouds of water vapor. The presence of water also dictates geological change, for Earth's oceans and atmosphere continually interact, creating weather systems and erosional forces that have reworked the surface of our home planet over the past 4.5 billion years.

LARGEST AND MOST FAMILIAR OF THE TERRESTRIAL PLANETS, the bluish-white sphere we call Earth is totally unlike anything else we can see in the universe. For one thing, it is the only place where we are sure life has arisen. It's also the only planet in the solar system whose surface still is constantly changing and developing.

The story of why Earth is so different from steamy Venus and arid Mars begins with the period of the great bombardment, when the debris of the solar disk was still thick around the planets. As each new object came streaking in, its mass was added to the Earth's growing accumulation and its enormous energy was converted into heat. Our home planet literally started to melt from the outside in, eventually becoming—like its sister planets—a glowing, molten ball in space.

During this molten period, minerals within the early Earth began to separate, much as salad dressing does when it's left standing. Heavier ones such as iron and nickel sank to the center, forming Earth's spherical core. Enormous internal pressures forced the inner portion of this core to become solid—a sphere of metal some 1,500 miles across. Farther out, where pressures weren't quite so high, the iron and nickel stayed liquid. (Motion within this liquid outer core continues to produce Earth's magnetic field.) Still farther out, lighter minerals containing silicates and oxides crystallized, forming the Earth's mantle. The lightest of all, such as quartz and feldspar, congregated atop the mantle much as frothy scum forms on a pot of boiling liquid. Rich in aluminum, calcium, and silicon, these minerals formed the thin outer shell we call Earth's crust.

This differentiation of heavy minerals from lighter ones resulted in a layered world, but like most natural processes, it wasn't perfect.

Here and there, traces of even the heaviest metals remained at the surface—and human beings have been mining them for millennia.

But, as with all other planets, only so much debris was nearby; eventually the rain of material from space slowed and the great bombardment ended. The rain never totally stopped, of course— every "shooting star" you see in the night sky is a leftover bit of that original disk of rocks surrounding the sun. As the number of impacts fell off, so did the influx of heat. Earth started to cool and solidify. Today, only its outer core remains liquid. It was during this period of cooling that a vital difference between our planet and other terrestrial planets began to show itself.

For deep within the Earth, locked into the rocks themselves, was an enormous reservoir of radioactive atoms. Some were the products of supernovae that had exploded billions of years earlier, some had resulted from the decay of primordial elements, still others were created in collisions of the solar wind with Earthly atoms. As time went on, the nuclei of these atoms, obeying the changeless laws of nuclear physics, underwent decay, converting a small part of their mass into a lot of heat. That process continues today.

It's not that radioactive decay didn't occur on other terrestrial planets—it did. It's just that smaller worlds such as Mercury, Mars, and our own moon had a lot less volume, therefore much less radioactive material to generate heat, and proportionally a lot more surface to allow that heat to escape. On these three bodies, heat generated by radioactivity—as well as heat that accumulated during the great bombardment—was able to leak quietly into space. The outer layers of these worlds froze, and any residual heat in their interiors could flow out through those solid layers by a process called conduction, similar to the way heat leaves a house in winter.

The larger terrestrial planets, Earth and Venus, contained too much heat to leak away solely by conduction, even over billions of years. On them, heat was brought to the surface by the movement of magma. This movement may have stopped long ago on Venus, but it continues on Earth. Our planet uses another process as well: plate tectonics, named for the fact that our planet's surface is broken into about a dozen separate plates. Apparently the Earth is the only planet in the solar system that removes its internal heat this way.

WITH EARTH'S RELATIVELY GREAT VOLUME AND SMALL surface area (compared to other terrestrial planets), the heat in its interior could not be entirely removed by conduction and volcanism. So the warm rock in its mantle began to rise to the surface, while cooler rock at the surface sank into the interior, a process that still continues. This ongoing exchange of hot and cool material between the Earth's deep mantle and its surface is a very efficient way for our planet to lose its heat. In a sense, the mantle started to boil like a pot of water on a stove, except that the time scale was very different; it takes mantle rocks hundreds of millions of years to "boil."

In response to this interior motion of the mantle, the planet's crustal layers also began to move, much as a thin film of oil does atop boiling water. Earth's lightest materials—the stuff that floated to the top during differentiation—became the granites and other rocks that make up the continents. Under these lay heavier rocks such as basalt, which we see occasionally when lava flows at the surface. It is these underlying rocks, which make up the ocean floor and the basement of the continents, that constitute the plates of Earth's surface. They, in turn, float on the "boiling" rocks of the mantle.

Sometimes, forces associated with plate tectonics tear plates apart, and we see a great indentation on the Earth's surface, hundreds of miles long. The Great Rift Valley that stretches from the Middle East down through Eastern Africa is one example (right). When continental plates collide, one gets pushed beneath the other, causing the continents they carry to crumple up into a mountain chain. The Himalaya were formed this way, as were Europe's Alps. When plates slide along each other, we get earthquake zones like the one associated with California's San Andreas Fault. The history of Earth's surface is really the story of the motion of its plates. They pull apart, push together, and move around in response to activity in the mantle beneath them.

This mobility of the plates has had some dramatic effects over geological time. For example, today we have a major ice cap at the South Pole, where ice more than a mile thick blankets the Antarctic continent. There is no corresponding ice cap at the North Pole because there is no continental mass; with no firm base to support it, a thick layer of ice can never build up. To have ice as massive as Antarctica's, you need a polar continent. But since Earth's continents move, this hasn't always been the case. Antarctica only arrived near its present location about 120 million years ago; the weather there got cold enough to form an ice cap just 36 million years ago. Before then, no ice cap existed. During Earth's long history there have been times when the planet was without major ice caps, times when it had two thick ice caps—when both poles were occupied by continents—and times, as now, when it had only one such cap. Each situation depends largely on the motion of the continents.

The Earth is still in the process of creation, still forming itself, a fact that holds interesting consequences for the way we view our planet. Nothing on Earth is permanent, not its mountains, oceans, or its deserts. The Appalachian Mountains, for example, took shape between 300 and 400 million years ago, a product of the Eurasian

DYNAMIC EVIDENCE

Two natural forces that continually reshape Earth appear in this view from Gemini 12, as clouds lined up by a 100-mile-per-hour jet stream bridge the Red Sea area, an active rift zone where two major tectonic plates are slowly rending the planet's surface.

South is to the top; the Nile River snakes through Egypt just right of center, while the tip of the Sinai Peninsula appears in the lower left corner. Far below the Red Sea's placid veneer, great tectonic forces continuously pull the shorelines apart, ever widening the rift. Eventually, the Middle East and Africa will separate.

plate repeatedly crashing into the North American plate, crumpling our continent's eastern seaboard into a range that once stood as tall and rugged as the Rockies are now. Today's Appalachians are worn down and smooth; in a few more hundred million years, they will be little more than hills.

EARTH'S FIRST ATMOSPHERE WAS FORMED FROM THE GASES emitted by volcanoes and magma outflows. But as soon as it formed, it began to change. The lightest molecules—hydrogen and helium, for example—started to leak into space, because the Earth's gravity simply wasn't strong enough to keep them home. It was, however, powerful enough to hang onto heavier molecules like water and nitrogen, so from very early on, Earth amassed a surrounding, protective layer of gas. Neighbors such as Mercury and our own moon were not so fortunate. Being so much smaller than the Earth—and having correspondingly smaller gravitational attraction—they soon lost all of whatever atmospheres they originally may have possessed.

As Earth cooled, the temperature of its atmosphere fell below 100°C and a very important event occurred—the first raindrop condensed. It was followed by many others, of course, as colossal rainstorms gave rise to Earth's oceans. Low areas flooded, and Earth became the blue planet that it has remained to this day. The oceans and rocks pulled carbon dioxide out of the atmosphere and absorbed it, lessening the greenhouse effect and allowing Earth's surface to cool even further.

In time, chemical reactions in our evolving atmosphere added simple organic molecules—the basic components from which more complex molecules of living systems are made. In addition, the debris that was still falling on the planet from space contained these and other molecules. One way or another, Earth's oceans soon contained the same basic materials. And then, by a process we still don't understand in complete detail, life began.

There are many theories about just how this occurred. One, the traditional scientific view, holds that tidal pools concentrated organic molecules as the pools periodically evaporated and were refilled by the oceans. Other theories, tied more closely to modern molecular biology, depend on chemical reactions driven by various molecules that might have developed. More recently, scientists have begun to examine the chemistry at the very bottom of the oceans, where titanic pressures and heat (supplied by upwelling magma) may have created conditions conducive to the formation of the first living things. At the moment, however, all we can say is that the origin of life on Earth—like the origin of the universe itself—remains a puzzle that modern science has not completely solved.

But somehow, that first living cell did appear on our planet. We know this happened relatively soon after the great bombardment and the formation of the oceans, because fossil evidence for the existence of fairly sophisticated living things has been found in oceanic rocks laid down off the coast of Australia some 3.5 billion years ago. We also believe that once life came into existence on the planet, it probably spread very quickly.

As life developed and branched out, it began to influence the planet itself. Some cells used the process of photosynthesis to convert sunlight and carbon dioxide into more complex molecules, giving off oxygen as a waste product. Oxygen, which is an extremely active element and was not present in Earth's first atmosphere, eventually built up to its present concentration of 21 percent. The extra energy available from oxygen-driven chemical reactions was exploited by plants, animals, and—ultimately—by human beings. The development of all Earth's varied life-forms took place in a constantly changing environment, where mountain chains were thrown up only to be weathered away, where seas were opened only to be closed. To survive, life had to change, adapt, and diversify, for it was filling niches that were ever in flux.

As we leave our home planet behind and head into the wider universe, perhaps we should remind ourselves just how dynamic and restless Earth is, this place where both the forms of life and the physical environment are constantly changing. Will we ever find another world like it anywhere in the universe? That's a question that has dogged humanity since time began.

YOU DON'T HAVE TO BE AN ASTRONOMER TO KNOW THAT Earth's moon varies in composition from one spot to another. Just look at the "man in the moon" on a clear night; the contrasting light and dark areas of the "face" are obviously different. The moon's dark regions were named maria (from the Latin for "oceans") by early astronomers, who compared them to Earth's oceans. Of course, we now know that no ships could sail on the lunar maria. For one thing, there's no air to fill their sails. For another, they're solid rock, the remains of ancient lava flows. Nevertheless, it is the maria whose dark expanses define the "face" we see in the moon. Lighter regions are the lunar highlands, which average about two miles higher than the maria. Most of the moon's craters occur in the highlands.

One of the most intriguing aspects of the moon for us Earth-dwellers is the fact that we always see the same side. Except for photographs taken by astronauts and various unmanned space missions, we've never seen its far side. This means that the time it takes the moon to turn on its axis—the lunar day—is exactly equal to the time it takes the moon to orbit Earth. That seems rather odd;

THE MOON

Once considered a cold and inert cinder, our cratered companion lacks a substantial atmosphere but is chemically akin to the Earth's outer layers. The moon is cooler and seismically less active than Earth; moonquakes are less frequent and weaker than earthquakes. Vast dark areas—the basaltic maria, or lunar "seas"—fill huge depressions in the lighter-hued and more heavily cratered lunar highlands, or terrae.

Earth's day and year are very different from each other, for example. Why is the moon's timing so perfectly coordinated?

The answer lies with the tides. We know that the moon's gravity raises tides on Earth, but it is equally true that Earth's gravity spawns tides on the moon—even though the moon hasn't any oceans. These tides cause the solid moon to flex slightly into a football-like shape. Earth's gravity then twists that football over time, so that the rotation of the moon on its axis either slows down (if the lunar day is less than a month) or speeds up (if it is longer than a month). The net result is that the long axis of the "football" comes to point continuously at Earth. In the jargon of the astronomer, we say that the moon has been de-spun; since its day is precisely as long as its month, the same face always points toward Earth. This happened early in the history of the solar system, and in fact most other moons in our solar system have been de-spun in the same way.

Since the moon is significantly smaller than the Earth—it has less than one-eightieth of Earth's mass—it exerts a much smaller gravitational pull. That's why, if you stand on its surface, you would weigh only one-sixth your normal weight. The moon's gravity is much too small to hold an atmosphere, so we find there only traces of neon, hydrogen, and helium. This means that there is no lunar weather to erode the surface features of the moon—although they are altered by the impacts of high-speed meteorites.

Like Mercury, our moon shows scars of impacts that date back billions of years. Its maria, in fact, are the remains of huge craters carved into the lunar surface during the great bombardment, around four billion years ago. Great lava flows followed, breaking through the weakened crust of the craters and filling them with molten rock, which quickly solidified. The mountainous regions around the maria actually are the rims of these ancient craters. Early astronomers interpreted them as ordinary mountains, and their notion that the lunar landscape was similar to the Earth's undoubtedly contributed to the idea that life might have existed there as well.

TO MODERN ASTRONOMERS, THE REAL IMPORTANCE OF THE moon lies in the fact that human beings not only have visited its surface but also brought back lunar rocks for study. Between July 20, 1969—when astronaut Neil Armstrong first set foot on the moon— and the final Apollo mission in 1972, about 700 pounds of rock and soil samples were brought to Earth. Such "moon rocks" are more than museum curiosities—they contain information not just about the moon but also about the formation of our solar system. Their chemical composition enables us to determine their age, through a technique similar to carbon-14 dating. So far, the oldest rocks found in the lunar highlands date to 4.4 billion years ago, the time when the

ENDURING MEMENTO

Telltale footprint left in the Sea of Tranquility by the Apollo 11 crew more than 30 years ago is destined to last a long time. Lacking an atmosphere, the moon has no weather, thus no weathering or appreciable erosion.

The only way to erase such marks of man would be for cosmic debris to rain down from outer space. Another possibility: Launch another team of astronauts to make a lunar cleanup.

moon first formed. The youngest moon rocks (found in the maria) are 3.1 billion years old and represent the last great outpourings of lava on the surface. Since then, our orbiting sister has been relatively quiescent, slowly accumulating craters.

In addition to collecting rock samples, the Apollo astronauts installed seismographs on the moon to monitor "moonquakes." For eight years—until NASA shut them down as an economy measure in 1977—these instruments told us about seismic waves traveling through the moon's interior. From this and other information, scientists pieced together a picture of the inside of the moon: Like Earth, it has a crust of light material. However, that crust is twice as thick on the far side as it is on the side facing us. This helps explain why there are so few maria on the far side, as impacts there didn't weaken the thicker crust enough to let lava through. Below this variable crust lies a denser mantle surrounding an even denser core, which may still be molten.

The average density of the moon is only about 3.3 times that of water, while Earth's is 5.5. Scientists puzzled a long time over how a body located so close to Earth could be so different. Today, many theorize that the moon formed in the aftermath of an event they call the "big splash."

Here's what they think happened: Once the Earth heated up and underwent differentiation, a lot of the heavy materials like iron and nickel became concentrated in its core, while the mantle was at about its present density. At that point, our planet was struck by an object perhaps as big as Mars; the collision created a huge crater and excavated large amounts of material from Earth's outer regions. Debris containing material from both the Earth and the colliding object splashed into orbit around our planet, and the moon subsequently formed there, in much the same way that the planets themselves had coalesced from the solar disk much earlier. Since this debris included parts of the Earth from which heavy elements already had been removed by differentiation, the moon's density was relatively low. In fact, its 3.3 density is about the same as that of Earth's mantle.

We believe that the moon accreted very rapidly—within a few hundred thousand years, just a microsecond geologically speaking. Once formed, it was heated by impacts and radioactive decay; its surface melted down to a depth of at least a few hundred miles, perhaps even to its center. By the time it was 200 million years old, that surface had cooled sufficiently for the highland rocks to solidify. The moon's most active geological period was from about 4.5 to 3.0 billion years ago, a span that complements our record of geological activity on Earth. In a sense, the moon is a fossil from the early solar system, still holding clues to our own planet's early history.

MARS, THE RED PLANET, NAMED FOR THE BLOODY ROMAN GOD of war, has long fascinated human beings. Today, we know that its red color—apparent even to the naked eye—is due not to blood but to the presence of iron oxides, or rust, on this planet's surface. Mars has long played a leading role in human mythologies, from various warlike deities to threatening creatures of science fiction. In 1938, when Orson Welles broadcast "War of the Worlds," based on the H. G. Wells classic tale about invaders from Mars, many in the radio audience mistakenly believed they were hearing a news report; their calls flooded emergency phone lines. This is only one instance of writers finding extraterrestrial pay dirt on Mars. It's an obvious choice, since the red planet is one of our nearest celestial neighbors.

One aspect of Mars we should dispose of right away concerns its "canals." Late in the last century, the prevailing theory of planetary development held that each of the four terrestrial planets represented a different stage, with those nearer the sun being younger than those farther out. In this scheme, Earth was a mature planet in the prime of life; Venus was still young and developing, while Mars had once been like Earth but was now the sad home of a dying civilization.

In 1877, Italian astronomer Giovanni Schiaparelli (whose family is now more famous for fashion than science) detected what looked like channels on the Martian surface. He called them *canali*—which can be translated to mean either natural channels or artificial canals, and some people chose the latter meaning. Given that humans were disposed to think of Mars as harboring life anyway, the conclusion that we had detected signs of civilization there quickly entered the popular consciousness.

The person most associated with Martian canals is the American astronomer Percival Lowell, who founded the Lowell Observatory in Flagstaff, Arizona. Looking at the red planet through his telescope, Lowell drew detailed maps of the fuzzy markings he interpreted as canals. He even calculated—to three decimal places—the rate at which vegetation marched southward during the Martian spring.

Yet the simple truth is that Mars has no canals of any kind. What Lowell and Schiaparelli saw on the Martian surface were blurry patches, the by-products of the best resolution that telescopes of their times could produce. The human mind often creates patterns (just look at psychology's Rorschach inkblot test), and this seems to be our best current explanation for the so-called canals of Mars. Too bad—they make great science fiction.

Today, we know that Mars is a most inhospitable place. Because its mass is only about one-tenth that of Earth's, most of whatever atmosphere it once may have had wandered into space billions of years ago. What remains today is thin and diffuse, similar in density to Earth's atmosphere at an altitude of 150,000 feet, roughly half as

MARS

More diverse than Mercury or our moon, the red planet contains heavily cratered ancient crust, younger plains, volcanic flows, canyons, and numerous channels reminiscent of floods on Earth. Huge dust storms rake its surface regularly, while at its poles, deposits of carbon-dioxide ice and water ice seasonally wax and wane. Mars also exhibits evidence that its crust once contained ice or liquid water—indeed, it may have some still.

VIKING 2

Setting up shop in 1976 in an area known as Utopia Planitia, the mission's lander takes in the Martian horizon as it extends a white boom to sample local weather. The rocky litter probably originated as ejecta from a nearby impact crater.

Mars owes its blood-red color to oxidized iron; its tawny sky stems from suspended dust in its atmosphere of carbon dioxide.

high as the space shuttle's orbit. Today's Martian atmosphere—what's left of it—is composed almost entirely of carbon dioxide.

Mars also is a cold world. There are a few spots on its equator where temperatures may creep above freezing during a summer noon, although any water that might form from melting ice would evaporate instantly, due to this planet's low atmospheric pressure. Over much of its surface, average temperatures hover around -60°C, while the polar regions can dip as low as -123°C in wintertime. There are polar ice caps on Mars, which once figured prominently in the story of the canals that supposedly conveyed polar waters to lower latitudes. These caps, however, are frozen carbon dioxide—dry ice—underlain by a residual cap of water ice. Enormous "snowfalls" of dry ice occur at the poles in winter, building up the ice caps, which evaporate the following summer.

Seen from Mars, the distant, faint sun punctuates a sky the color of butterscotch—thanks to fine dust particles suspended in the Martian atmosphere, which selectively scatter light. The planet's surface is mostly rock and sand, laced with a few striking geological features—a large but probably extinct shield volcano that stands

higher than Mount Everest, and a vast nearby valley named Valles Marineris. This valley is longer than North America is wide, and far deeper than the Grand Canyon. Overall, the history of Mars is similar to that of other small terrestrial planets—molten origins followed by rapid cooling and a cessation of geological activity.

Modern exploration of Mars began in 1965 with Mariner 4, the flyby that sent back the first crude photos of the Martian surface. In 1976, the two Viking spacecraft arrived at Mars. While the orbiters mapped the planet in unprecedented detail, the landers took pictures from the surface. This mission redefined our concept of the red planet.

You have to remember that back then, the burning question about Mars didn't concern its geology or current atmosphere, but whether or not it harbored living things. The sole purpose of several experiments on the Viking landers was to search for chemical evidence of life in the Martian air and soil. In one of them, water and nutrients were added to soil, and scientists waited to see if anything grew. Another experiment used compounds that had been tagged with radioactive nuclei, to find out if some life-form would process them and perhaps release some of those tagged nuclei to the atmosphere. By the end of the mission, scientific consensus held that there was no widespread life on Mars, although a few researchers tried to figure out ways that living forms might exist in odd corners of the planet and so could escape detection by Viking. One popular theory suggested that microbes survived beneath the surface, perhaps in a dormant state, in what seems to be the Martian equivalent of permafrost.

But if Viking's landers diminished human hopes for finding life on Mars, the orbiting sister ships confirmed results that had been suggested by earlier Mariner flybys. Looking down on the red planet, they sent back detailed pictures of geological formations that could have been created only by running water. Long channels, looking for all the world like terrestrial riverbeds and deltas, provided evidence that even though Mars might be dry and dusty today, there was a time when water flowed freely on its surface. Some experts even suggested that geological formations in the northern hemisphere showed that at one time Mars may have had an ocean. The channels sighted by Viking, by the way, were definitely not Schiaparelli's canali—they are much too small to be seen from Earth, even with the most powerful telescopes.

The current picture of Mars that is emerging differs both from the "ancient life" idea espoused at the beginning of the 20th century, and the "permanently freeze-dried" image that followed. The dominant theory today holds that sometime during the early development of Mars—perhaps the first billion years or so—it and Earth were not very different. Surface temperatures were high enough to maintain considerable bodies of liquid water on both planets, and both had atmospheres. It was only later that, as the Martian atmosphere leaked slowly into space, the red planet began to assume the barren form we now see.

This image of a wet early Mars was dramatically buttressed on July 4, 1997, when the Mars Pathfinder landed on the red planet. Wrapped in a giant air balloon that allowed it to bounce repeatedly without breakage, Pathfinder came to rest in the middle of what seems to be an ancient floodplain, where massive outwash from a huge lake once carved the surface. Indeed, the spacecraft's striking photographs showed a jumbled surface full of boulders, rocks, and cobbles very similar to flood-ravaged regions on Earth. Even the Twin Peaks, two low hills near the lander, appear streamlined, as if shaped by running water. NASA now plans to send more missions to Mars in the near future—one approximately every two years, in fact, while the planet is relatively near Earth. In 2008, with the help of an orbiter

SCIENCE FICTION...

Writers have had a field day with Mars, repeatedly making it the home base of extraterrestrial beings in tales such as H. G. Wells's *The War of the Worlds* and the Hollywood spoof *Mars Attacks!*

It's an obvious choice: Mars is right next door, with surface features visible from Earth (thanks to the thin Martian atmosphere). Also, its length of day, angle of its tilt, and its seasonal changes are strikingly similar to Earth's.

launched by CNES, the French space agency, NASA hopes to bring back some Martian rock samples for study.

If Mars and Earth really were that similar billions of years ago, another intriguing hypothesis presents itself. We know that life developed fairly quickly here, producing complex ecosystems within 500 million years of the end of the great bombardment. If there had been oceans, ponds, or even ice-covered lakes on Mars at that time, it's reasonable to suppose that life may have developed there as well. It may only have consisted of single-celled organisms that went extinct whenever the surface waters disappeared, but while such life forms lasted they would have left their mark on Martian rocks, in the same way that bacteria have on our own planet. Scientists have been pondering how to detect fossils in Martian rocks, and hopefully will be ready when the first samples reach Earth.

Some researchers, not content to wait until 2008, have begun to search for life in one of the most intriguing resources already on Earth—the dozen or so meteorites that apparently came from the red planet. They believe these rocks once rested on the Martian surface but were blasted off by some meteoritic impact, then wandered in space for a time before falling to Earth as small meteorites themselves. Scientists feel certain these rocks are from Mars because tiny bubbles of gas trapped inside the rock perfectly match the composition of the Martian atmosphere.

In 1996 one group of researchers examined a meteorite with high-powered microscopes operating at the edge of their capability, and claimed to find evidence of tiny fossils. Others disputed their findings, arguing that the structures detected were ordinary mineral formations, not the remains of once-living things.

But even though Mars clearly lacks the kinds of intelligent life we once hoped to find there, it may very well contain evidence for life's failed experiments back in the early days of our solar system. As such, it could tell us a lot about how life might erupt elsewhere in the universe, a subject to which we'll return.

UNLIKE MERCURY OR VENUS OR, FOR THAT MATTER, EARTH, Mars has two moons, Phobos (Fear) and Deimos (Terror), named for the mythological attendants of Ares, the Greek god of war. They are lumpy, irregularly shaped rocks whose longest dimensions are less than 20 miles, and whose makeup seems entirely unlike that of their mother planet. Phobos orbits about 3,700 miles above the Martian surface, so near that an astronaut standing on it would see the dazzling red face of Mars fill nearly half the sky. Deimos is much farther out, averaging an orbital distance of some 14,500 miles and exhibiting a somewhat smoother appearance. Both are thought to be asteroids that were captured by Mars's gravitational field.

...AND SCIENCE FACT

Believed to have been blasted from the surface of Mars by some titanic impact 16 million years ago, this 4.5 billion year-old fragment of a meteorite—known as ALH 84001—fell to Earth about 13,000 years ago. It was found in Antarctica in 1984.

Carbonate globules and other inclusions within its cracks, some scientists believe, may be the work of primitive, bacteria-like growths that must have lived on Mars. Others disagree, but the search for Martian life continues.

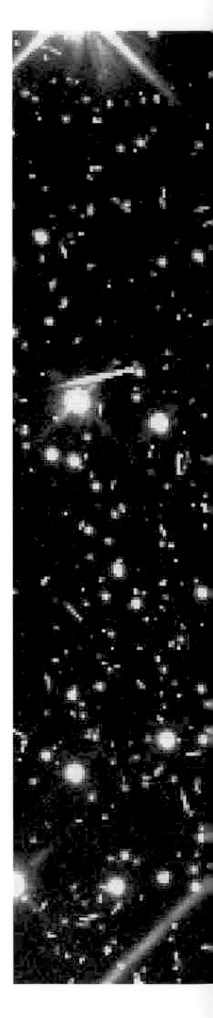

THE FINAL STOP ON OUR TOUR OF THE INNER SOLAR SYSTEM lies well beyond the orbit of Mars, in a planetless region known as the asteroid belt. Here, millions of asteroids—chunks of rock similar in size and shape to Phobos and Deimos—orbit the sun in a wide swath. Astronomers discovered the first asteroid in 1801, and have since located and named several thousand. In October 1991, while journeying toward Jupiter, the Galileo spacecraft secured our first close-up views of individual asteroids. Only a half dozen of these rocky objects span more than 200 miles; Ceres, the largest asteroid yet discovered, is approximately 600 miles across—roughly the distance from Philadelphia to Detroit. A much more modestly sized one, Geographos, was named for the National Geographic Society, which has chronicled space exploration for decades. Yet another asteroid, Ida, is about 35 miles by 9 miles—and it boasts an even tinier moon, named Dactyl.

Scientists once speculated that the asteroid belt might be the remains of an exploded planet, but by the 20th century they realized that this belt doesn't contain enough material to make a full-fledged planet. (The estimated total mass of all asteroids combined is only one-twentieth of the mass of our moon.) Currently, we believe that asteroids are debris left from the solar system's formation—perhaps kept from coalescing by the gravitational pull of Jupiter, perhaps formed by a more complex process. One proposal suggests that the asteroid belt once held several Mars-size objects, before collisions shattered them and ejected most of their bulk into space. In this theory, the red planet is the only one of those objects left in the solar system. Whatever the final explanation of the asteroid belt, it seems clear that Jupiter's gravitation has played an important role.

Occasionally, through the influence of Jupiter or of collisions, asteroids are thrown into orbits that bring them closer to the sun and inner planets. During the past decade, one came within about 106,000 miles of Earth, less than half the distance to the moon. Such so-called Earth-approaching asteroids might actually collide with the Earth some day, prompting concern that they could cause consequences as grave as those that accompanied the extinction of dinosaurs 65 million years ago. So it is that scientific surveillance of asteroids has increased. In 1996, NASA launched NEAR, the Near Earth Asteroid Rendezvous spacecraft, which should soon go into orbit a mere 20 to 30 miles from the asteroid Eros. This asteroid, about 20 miles long in its greatest dimension, will be studied intensively for about a year.

Moving ever outward from the sun, we leave the more familiar realm of the terrestrial planets, passing even the outermost asteroid in the asteroid belt. Ahead lies a very different region, the realm of the outer planets, also known as the gas giants. ●

ASTEROID BELT

Between the orbits of Mars and Jupiter lies the asteroid belt, a realm devoid of planets but populated by millions of rocky residents that vary greatly both in shape and size.

One named Mathilde (above) measures about 40 miles across and bears the scars of numerous collisions. A newly discovered asteroid traces a blue arc (right) in an image from the Hubble Space Telescope. The finding and cataloging of asteroids remains an ongoing scientific effort.

MERCURY

Named for the winged messenger of the gods, Mercury takes only
88 Earthly days to sweep around our sun. It remains the most difficult
inner planet to study, due to the sun's ever present glare. This dramatic
photomosaic represents a recent reprocessing of Mariner 10 data,
gathered during three flybys of the planet in 1974 and 1975, which
revealed surface features as small as a mile across. Smooth patches
reflect areas where no data exist. Mercury's pocked surface seems
to show more extensive volcanism than does our moon, with plains
nearly as reflective as its cratered regions.

TESTAMENT OF TIME

Ravaged wedge of Mercury's face
(right) bears the scars of thousands
of ancient impacts, many dating
to the great bombardment that
occurred early in the solar system's
development. One close-up
(opposite) reveals a tiny, relatively
recent crater nicking the rim of the
much larger Brahms crater, which
encircles a central spike of rock.
By studying overlapping features
such as these, scientists are able to
deduce much of the planet's history.

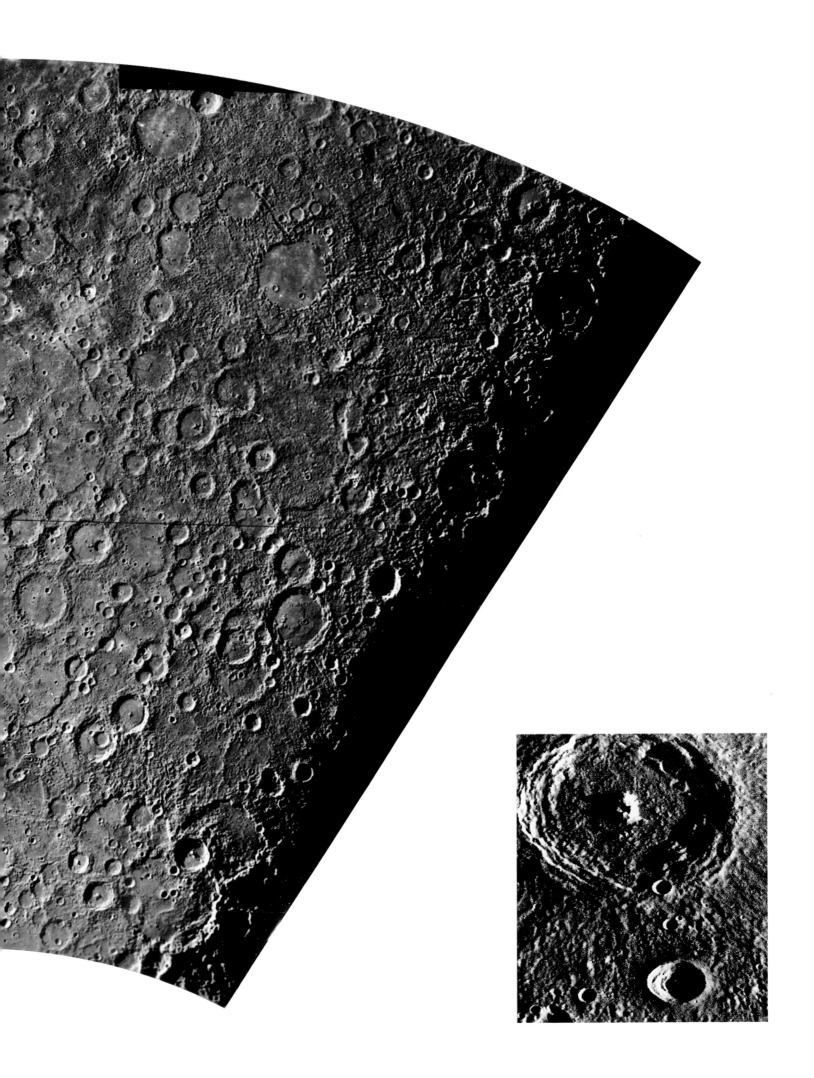

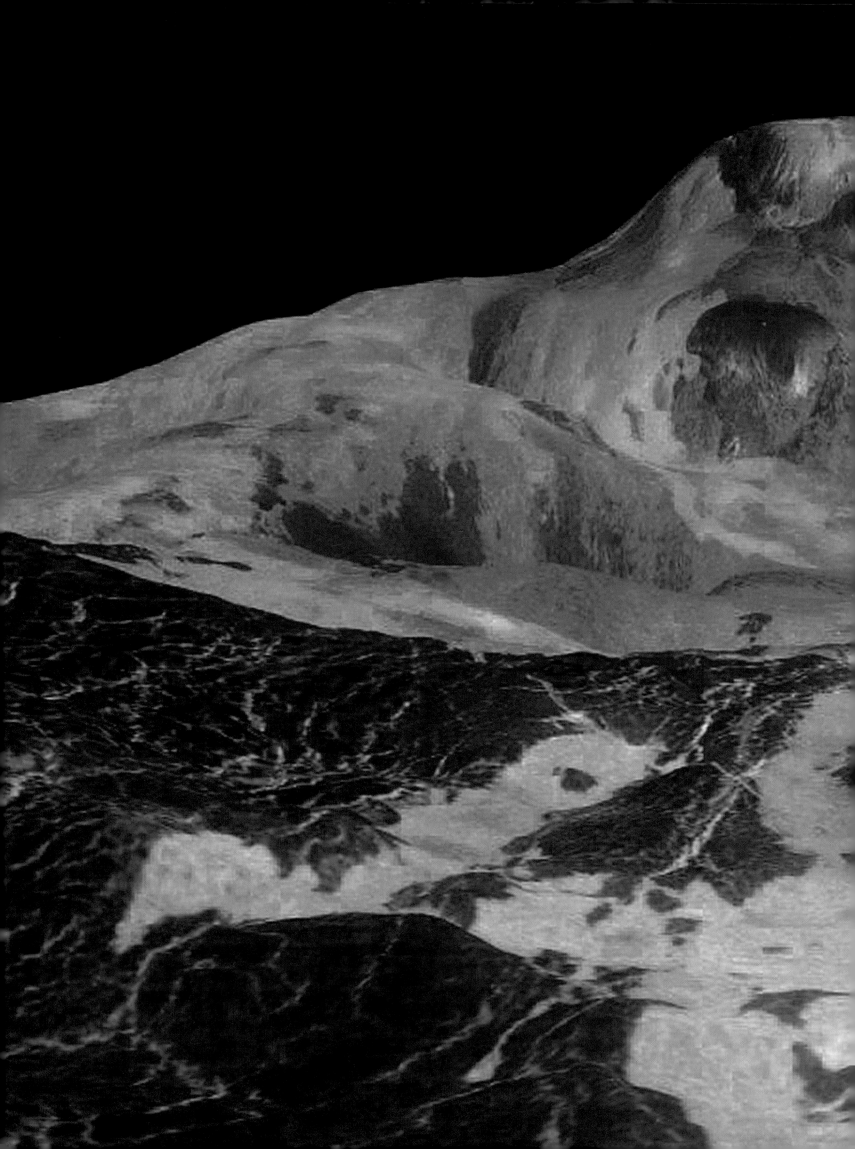

Namesake of the ancient Egyptian goddess of justice and truth, Maat Mons towers five miles high amid hellish surroundings. Surface temperatures on Venus hover between 450°C and 480°C day and night—making this planet hotter, most of the time, than Mercury—while droplets descend from dense sulfuric acid clouds. This three-dimensional, computer-generated image was created with data from Magellan, which employed powerful radar bursts to "see" through the normally impenetrable Venusian atmosphere. Colors reflect actual hues, as recorded by the Soviet Venera probes; the vertical scale has been exaggerated to emphasize differences in terrain.

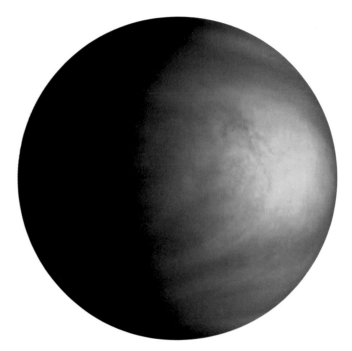

WITH CLOUDS—
AND WITHOUT

False-color view of Venus (above) from the Galileo spacecraft casts the planet in blue to emphasize subtle contrasts in its clouds. Such techniques have enabled scientists to determine that Venus is prone to winds that flow east to west, at around 230 miles per hour.

A very different vision of Venus (left), assembled with the help of Magellan's cloud-piercing radar, shows how the planet would appear to us if there were no dense cloud cover. Colors, based on images from the Venera spacecraft, approximate actual hues.

360-DEGREE VIEW

Packing both hemispheres of a round Venus into a single flat image, this sinusoidal projection illustrates the planet's varied topography across its entire surface. The prime meridian and each parallel reflect actual distances, while other longitudinal lines are distorted into sine curves.

Such projections, originated in 1606 by Flemish cartographer Gerardus Mercator, are particularly suitable for comparing areas of different sectors. Colors show relative elevations, with reds identifying the highest, followed by yellows and greens; blues mark the lowest regions.

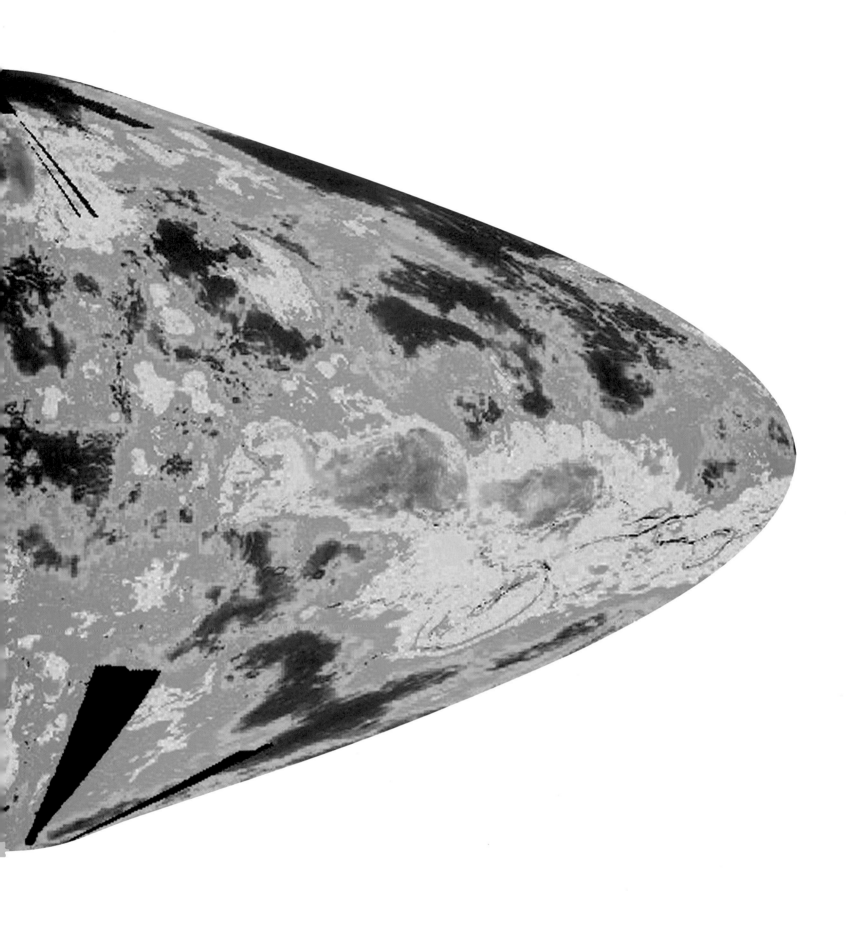

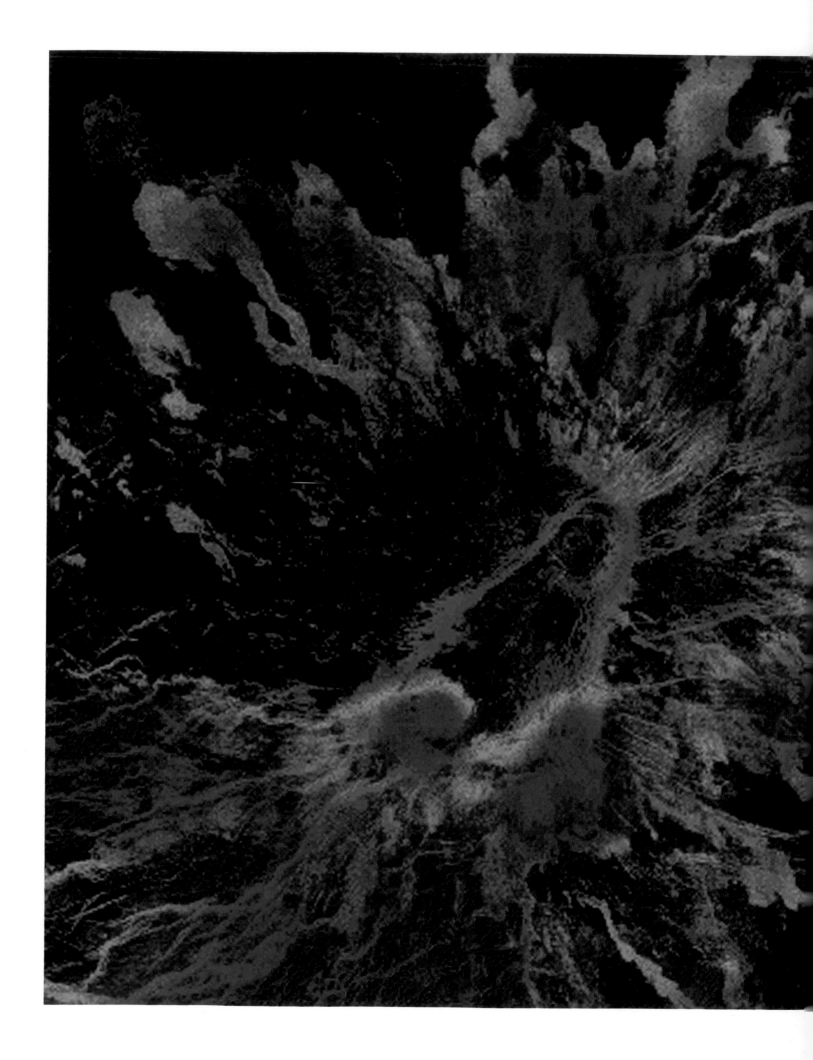

VENUSIAN VOLCANO

All but hidden by its own surreal
ejecta, a volcanic cone more than a
mile high studs the planet's southern
hemisphere in this colorized radar
image from the Magellan spacecraft,
which focuses on an area roughly
400 miles square. Red and orange
represent hot areas and old lava
flows, while the bluish center
signifies cooler temperatures at
the volcano's summit, possibly
indicating the presence of certain
iron-sulfur compounds that may be
unstable at lower altitudes.

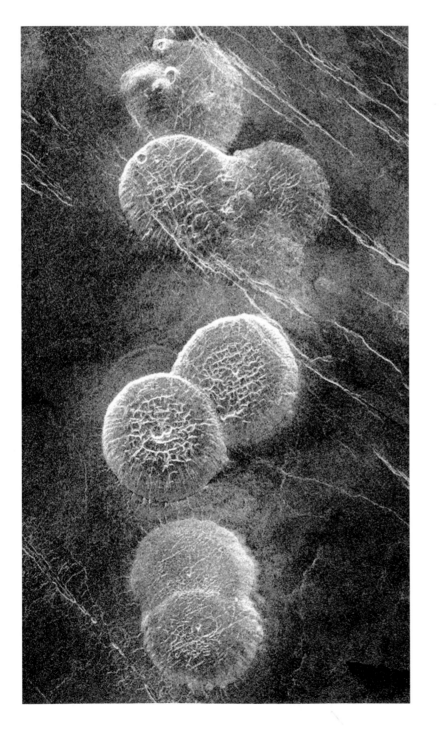

DETAILS OF VENUS

Evocative features abound on the Venusian surface, inspiring some fanciful labels. Magellan's all-seeing radar has produced numerous highly detailed pictures of Venus's widely varying topography, which scientists strive to explain by comparing to structures and processes on Earth. So-called pancake volcanoes (left) apparently result from eruptions of extremely thick lava.

One standout volcano, categorized as a "tick" for its arachnid shape (opposite), stretches more than 40 miles across, its large, concave summit ringed by radiating ridges and valleys. On one side, dark lava flows breach the volcano's rim, giving rise to this tick's "head."

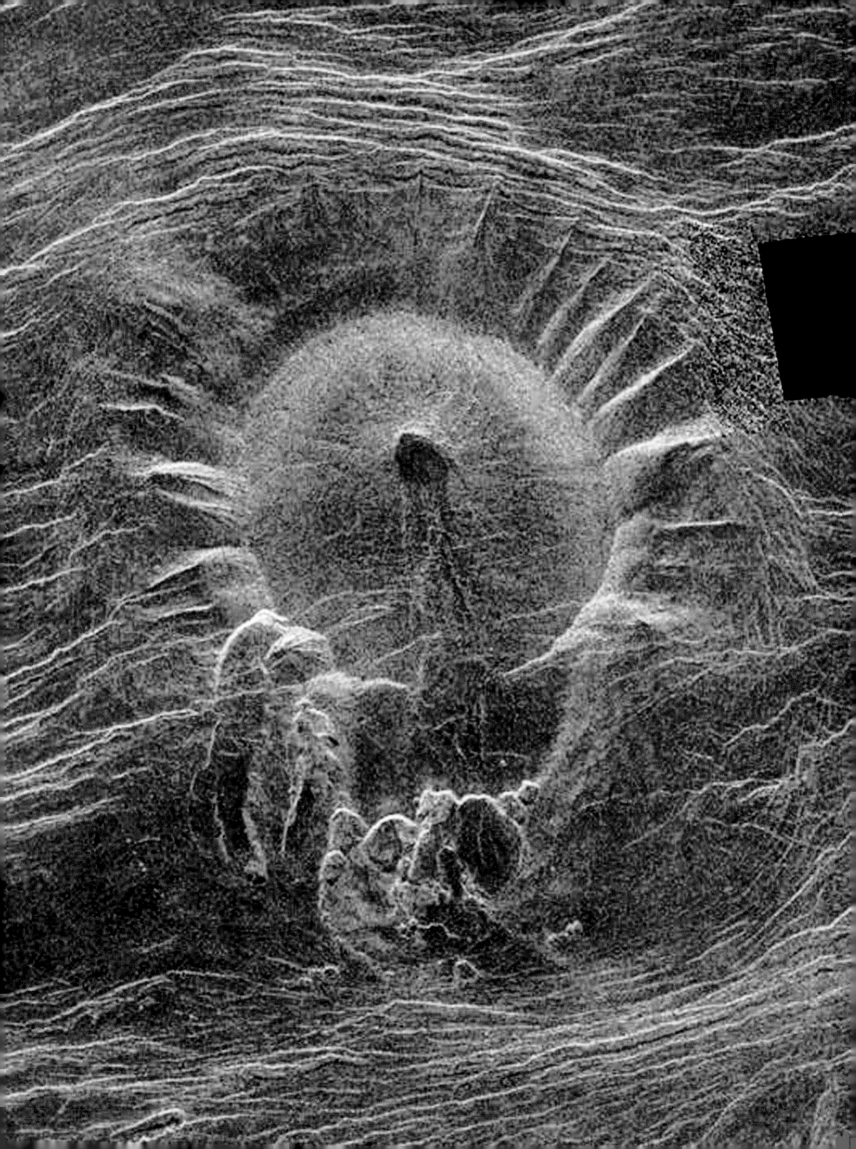

Viewed from the space shuttle *Discovery*, Earth's aurora australis glimmers
in the upper atmosphere, its waves of green and red light hovering
between an overexposed moon and the cloud-dappled southern Indian
Ocean. Such displays result when high electromagnetic activity on the sun
increases the solar wind, which collides with the Earth's magnetic field
and with the molecules of Earth's atmosphere, ionizing them.

SEEN FROM SPACE

All sorts of satellites equipped
with all sorts of instruments orbit
Earth today, monitoring a wealth of
different variables as they expand
our knowledge of the planet. Sensors
aboard the SeaStar spacecraft chart
subtle differences in the occurrence
of phytoplankton—the essential
starting point of all marine food
chains—by measuring differences
in reflectivity of the world's oceans,
both in the Northern (above) and
Southern (left) Hemispheres.

Other satellites record changes
taking place in the seasonal ozone
hole over Antarctica (right), which
has been growing larger, spurring
environmental concerns. In this
false-color image, low ozone levels
register as blue.

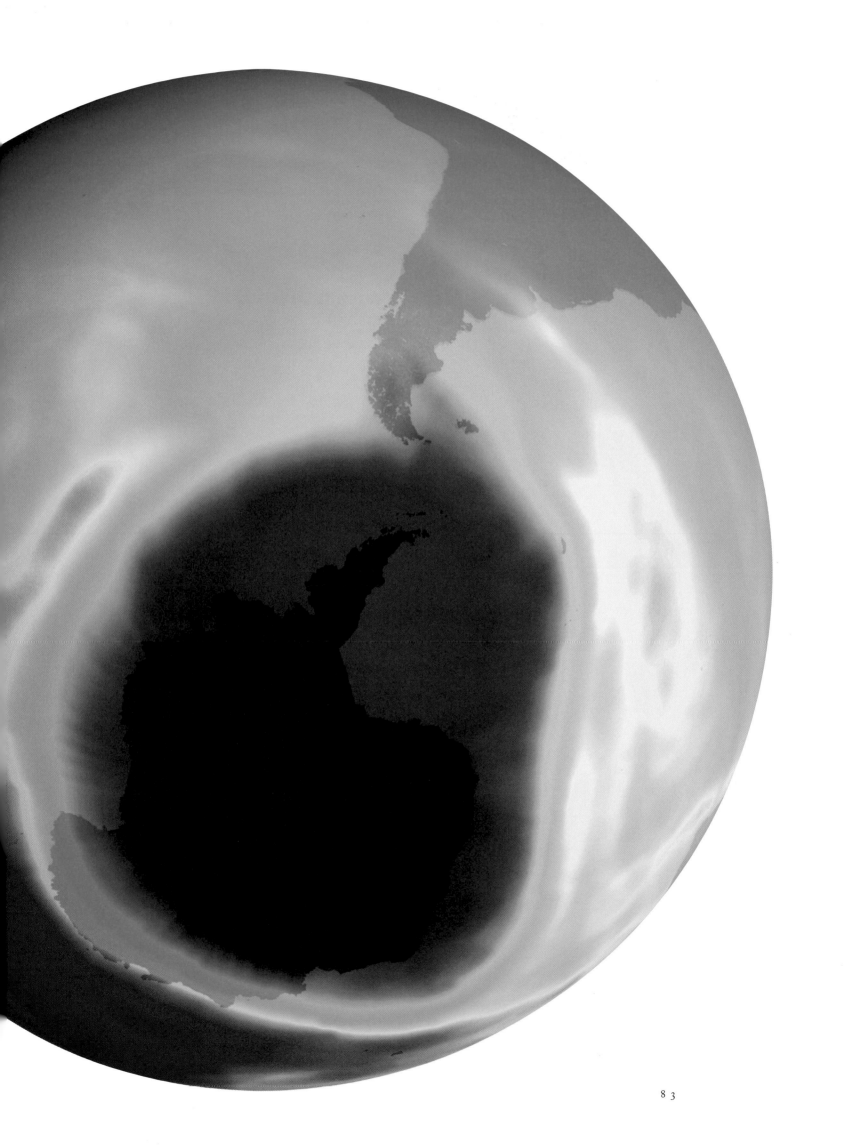

HIMALAYA

[PRECEDING PAGES] Gloriously tortured
topography of the Earth's highest
mountain range attests to the power
of tectonic forces far below our
planet's crust. This false-color radar
image—used by geologists to map
the distribution of various rock types
—helps them understand how
major geological features formed
and have been eroded. Granites
show up as brown-orange, while
older sedimentary and volcanic rocks
appear in pale blue on some ridges.

KAMCHATKA

Detailed false-color radar image of Siberia's Kamchatka Peninsula (above) highlights another widespread geologic reality on Earth: volcanism. Here, the volcano Maly Semlyachik, part of the Pacific Ocean's powerful Ring of Fire, rises near the coast. Long feared for their awesome power, volcanoes also constitute a major creative natural force on Earth, providing new crustal material, contributing to the chemical composition of the atmosphere, and concentrating various minerals and ores.

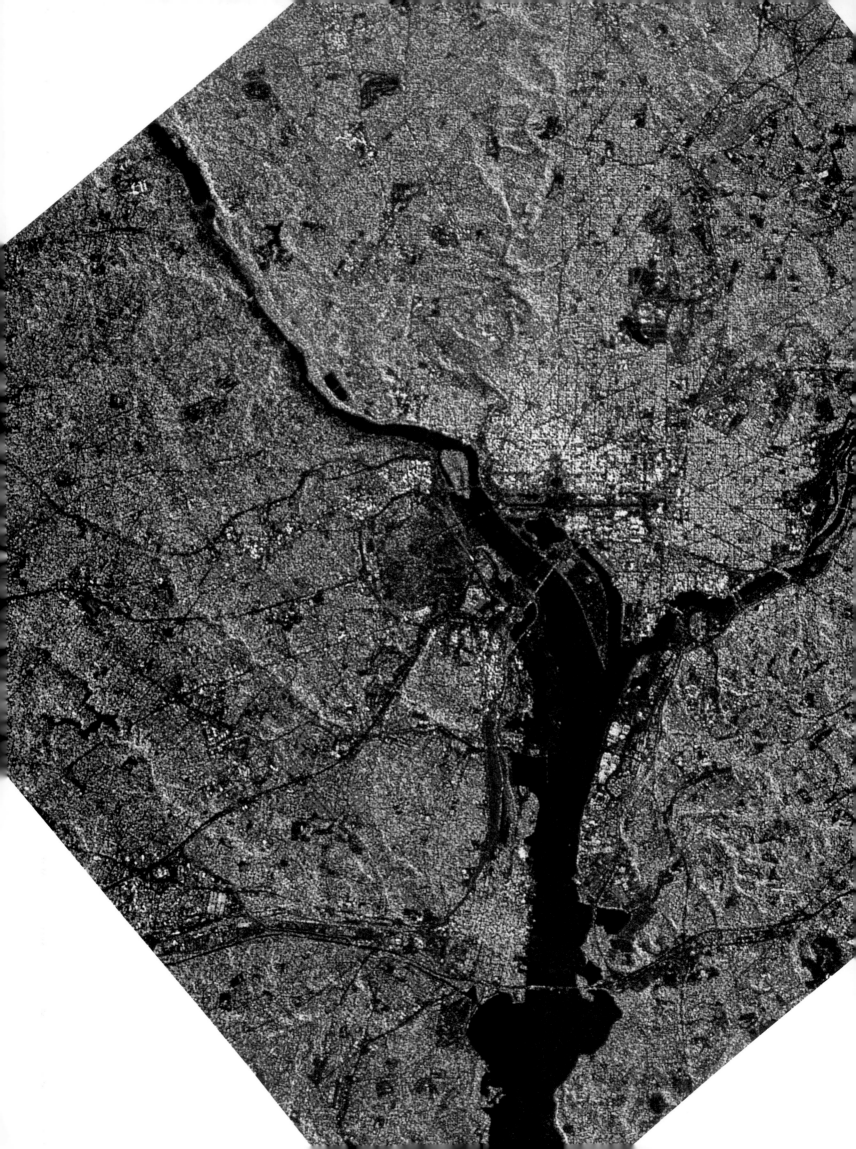

WHAT'S DOWN THERE?

In addition to mapping natural
topographic details across the face
of our and other planets, satellite
imagery also can monitor human
activities. This space-shuttle radar
image shows the Washington, D.C.,
area (left), depicting the most
urbanized regions in white and light
blue, vegetated areas in darker
blues, all set off by the black of the
Potomac and Anacostia Rivers.

   In contrast, a view from space
of the northern Rocky Mountains
(above) chronicles the dendritic
branching of ridgelines and valleys
along our Continental Divide.

LOOKING BACK

This false-color view of our home
planet comes to us from the
Clementine spacecraft, whose
primary mission was to explore
the moon. The bright arc just inside
Earth's blue envelope, called airglow,
marks a zone of faint luminescence
caused by chemical reactions in the
upper atmosphere. The hazy bright
spot on the right is due to the lights
of an urban area.

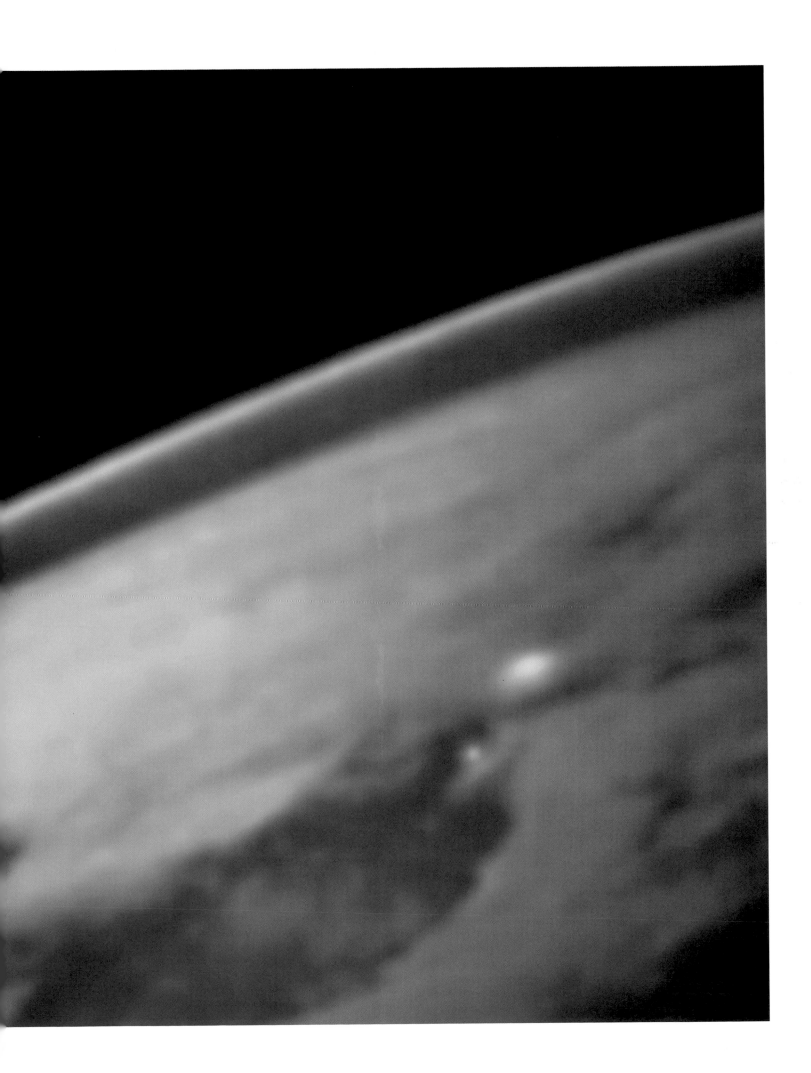

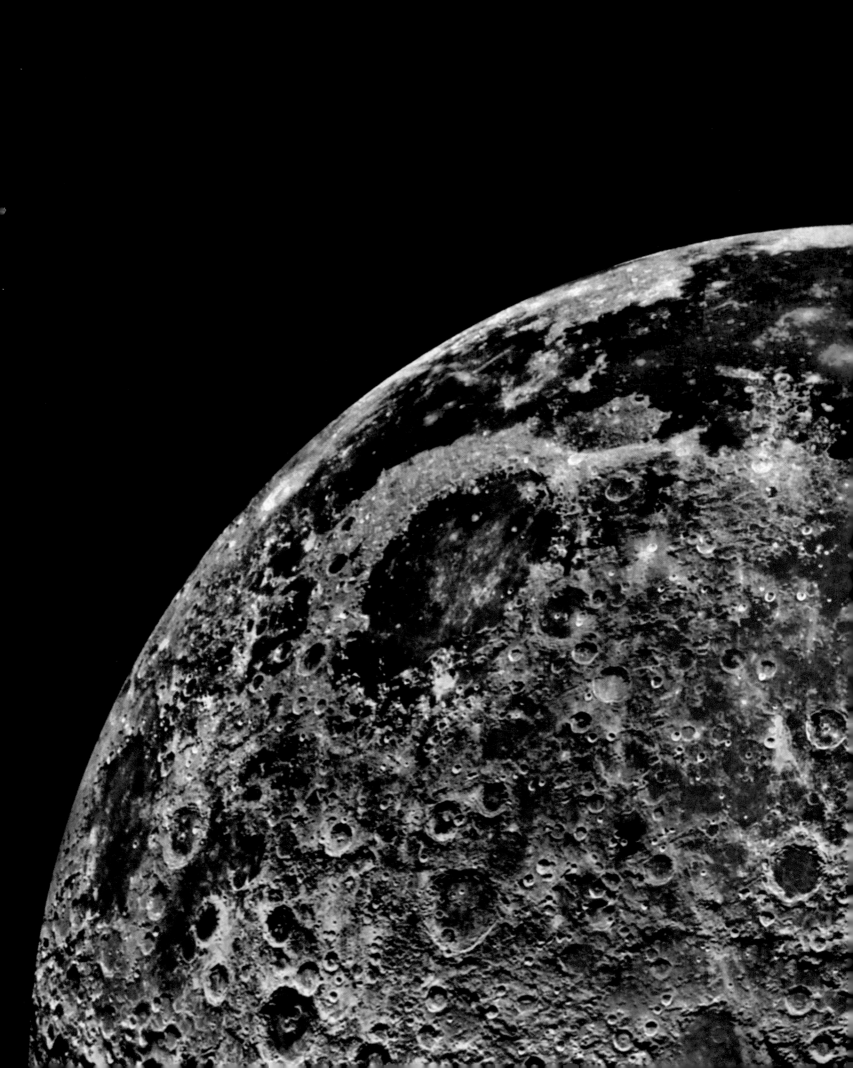

THE MOON

Resembling some fantastic dessert, this false-color mosaic maps the mineral composition of the moon's surface. Bright pink identifies lunar highland areas; blue and orange indicate volcanic lava flows and the lunar maria. Dark blue regions are high in titanium, while lighter blues signify other mineral-rich soils associated with relatively recent impacts.

ALL ALONE ON
THE LUNAR SEA

Scientist and astronaut Harrison
Schmitt, lunar module pilot for the
final Apollo mission, takes a turn
with the lunar rover near Apollo 17's
desolate landing site at Taurus-
Littrow impact valley. Long after
lava filled the valley some 3.7 billion
years ago, an avalanche brought
down the highland rock that still
litters this moonscape.

   Due to the moon's small size and
gravity, it lost whatever atmosphere
it once may have had to space eons
ago—so this and other bleak vistas
show no effects of wind, water, or
life. Orbital patterns dictate that,
each month, most sectors of the
moon spend about two Earthly
weeks in sunshine and two weeks
in darkness.

HALF AND HALF

Because gravity has locked the moon into synchronous orbit with Earth, one side continuously faces us. This Apollo 16 image includes part of the familiar, mare-splotched "man in the moon" (left) and part of the highly cratered far side (right). Scientists think the dearth of lunar maria on the far side indicates that the crust there is thicker, making it less likely for impacts to cause ruptures and magma flows.

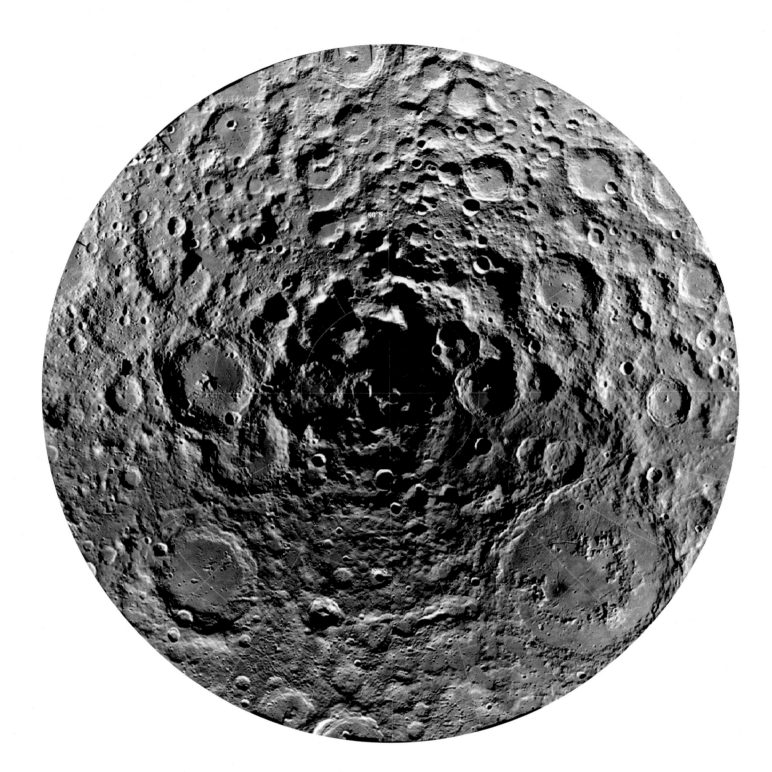

SOUTHERN EXPOSURE

This mosaic of the moon's south polar region contains data from some 1,500 Clementine images. A major depression surrounding the pole may be dark and cold enough to harbor traces of cometary ice.

One of the moon's youngest impact basins, Schrodinger, occurs near the edge of this image at about four o'clock; it includes one of the largest lunar cinder cones.

STARK RELIEF

Long shadows—due to the sun's low elevation and the harshness of sunlight in a world without an atmosphere—accentuate the topography of the lunar surface in these two views. Impact craters, younger ones dappling earlier ones (above), help scientists decipher some aspects of lunar geology. The orbiting lunar module from one Apollo mission sent back an image (right) of the crater Copernicus, backed by the Carpathian Mountains.

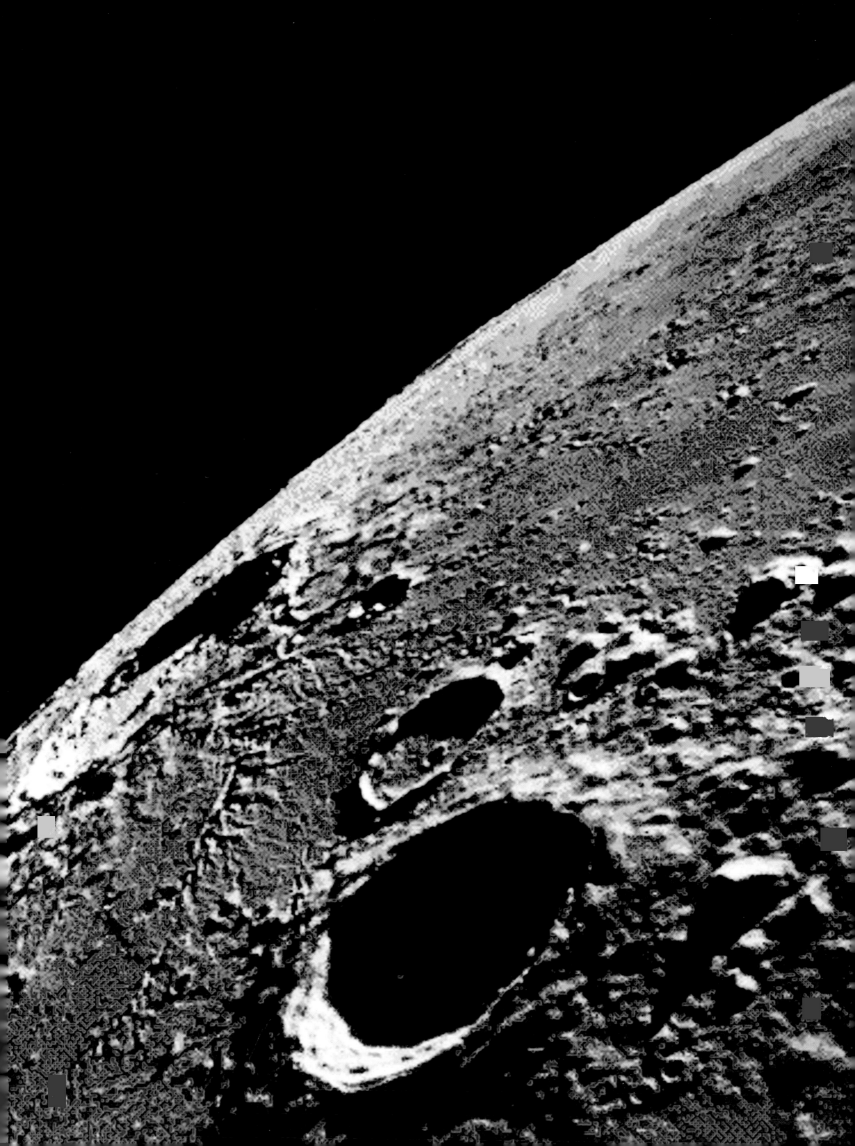

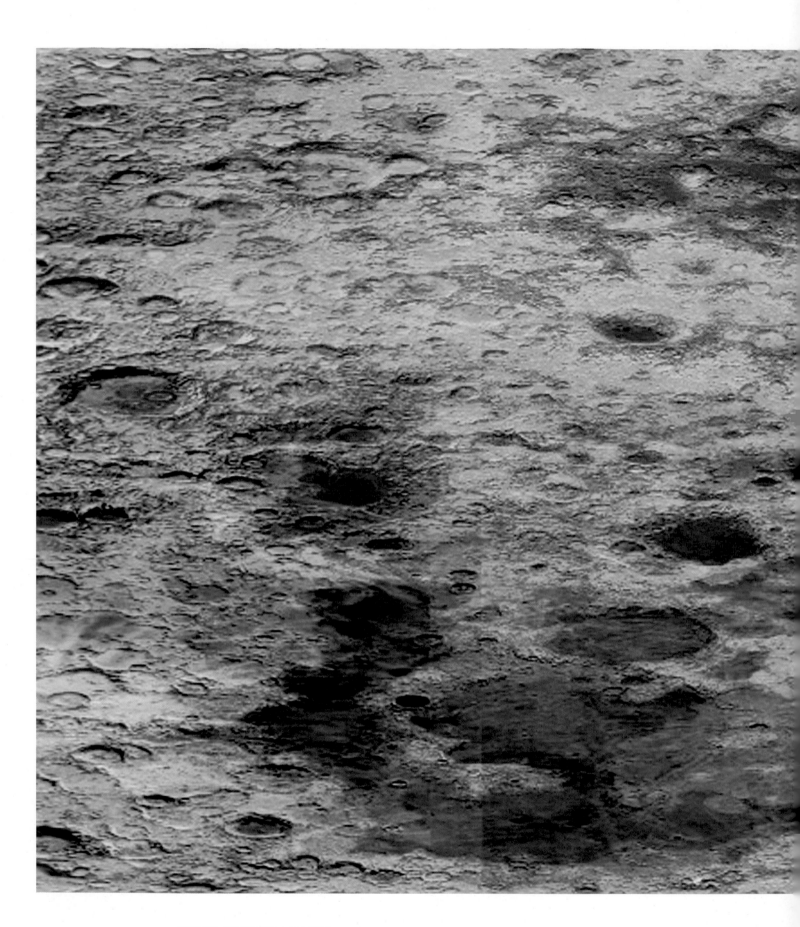

FALSE-COLOR MOON

Keyed to differences in elevation, this lunar portrait indicates the lowest areas in purple, with blue, green, and red signifying progressively higher altitudes. Lowest of all is the South Pole–Aitken Basin, at upper

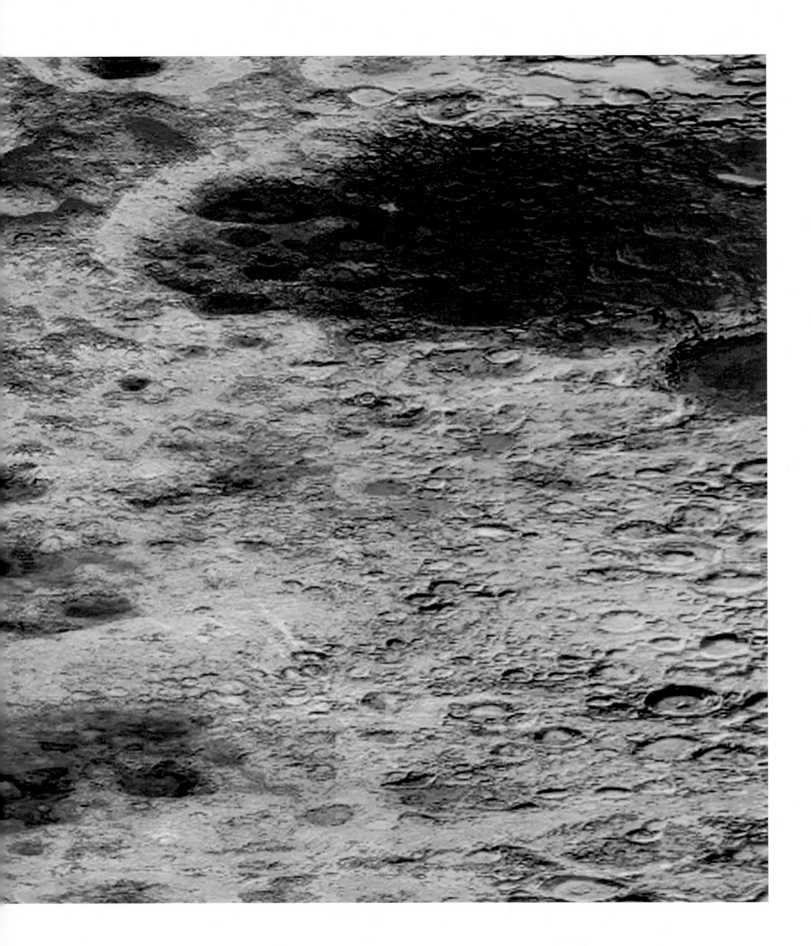

right. The other major depression
visible here may have been caused
by one or more impacts. Reddish

highlands at top center could
have been created by ejecta
from earlier impacts.

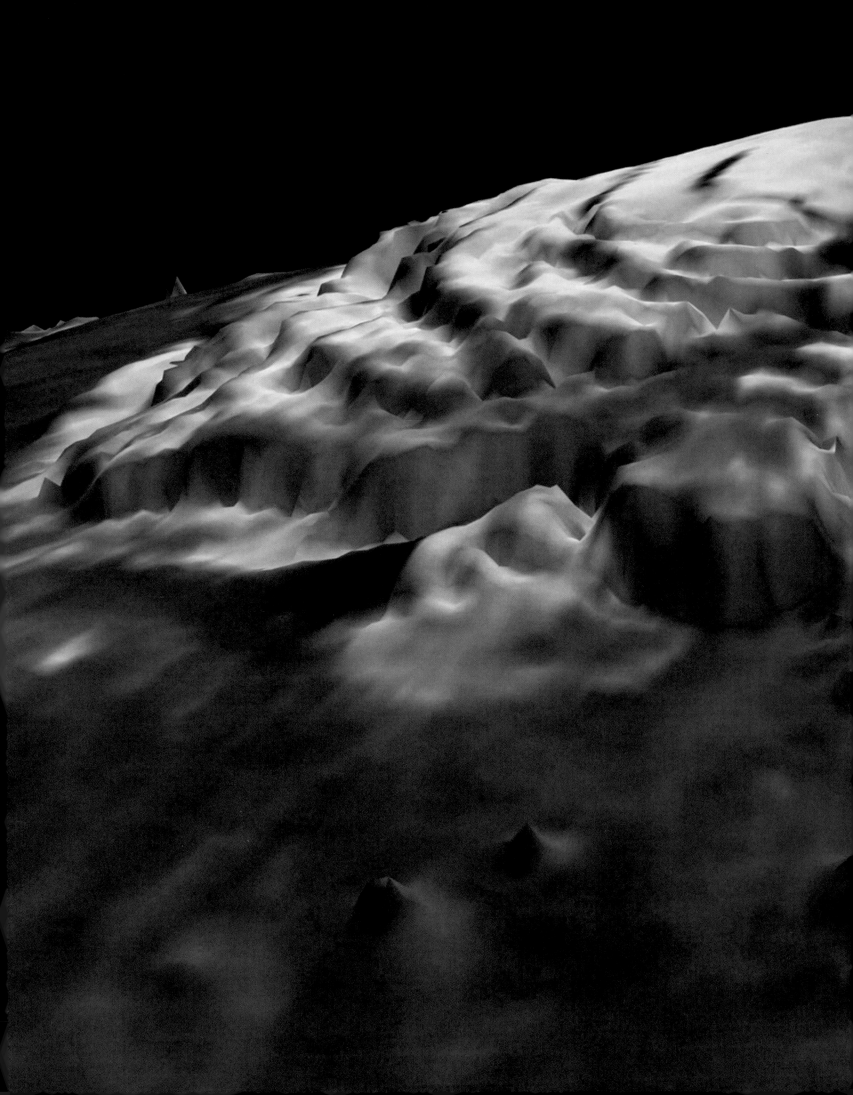

Laser altimetry measurements from the Mars Global Surveyor spacecraft helped create this image, the first three-dimensional view of the Martian north pole, in 1998. Such technological advances enabled scientists to improve estimates of the volume of the red planet's polar cap, and to study surface variations and the heights of its wispy cloud layers. The polar caps of Mars contain some water ice overlain by frozen carbon-dioxide ice.

FREEZE-DRIED PLANET

Mars (right) lies about 137 million miles from the sun—half again as far as the Earth—resulting in an average surface temperature of minus 63°C. Smaller than our planet, it has less than half Earth's surface gravity. Its towering volcanoes are extinct, and it lacks liquid water. An immense canyon system, Valles Marineris, gouges its midriff for some 2,500 miles, plunging as deep as 6 miles.

Its two small, irregularly shaped moons, Phobos (above) and Deimos (below), may be captured asteroids. Their names translate to Fear and Terror, memorializing the attendants of Ares, Greek god of war.

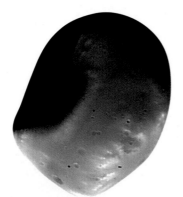

PATHFINDER ON SITE

This panorama from the Mars lander known as Pathfinder shows the twin ramps and cylindrical antenna, deployed amid its rocky touchdown site near an apparent outflow channel. Sojourner, the lander's six-wheeled rover, rolls up to the nearby rock NASA scientists dubbed Yogi. Launched in December 1996, this mission reached Mars on July 4, 1997, becoming the first to land on the red planet since Viking 2, more than 20 years earlier.

Pathfinder promptly—and very reliably—radioed back spectacular close-ups of the Martian surface that held Earthly audiences in thrall for weeks. One of its true-color images (opposite) shows a distant sun setting on a Martian landscape amid a dingy sky.

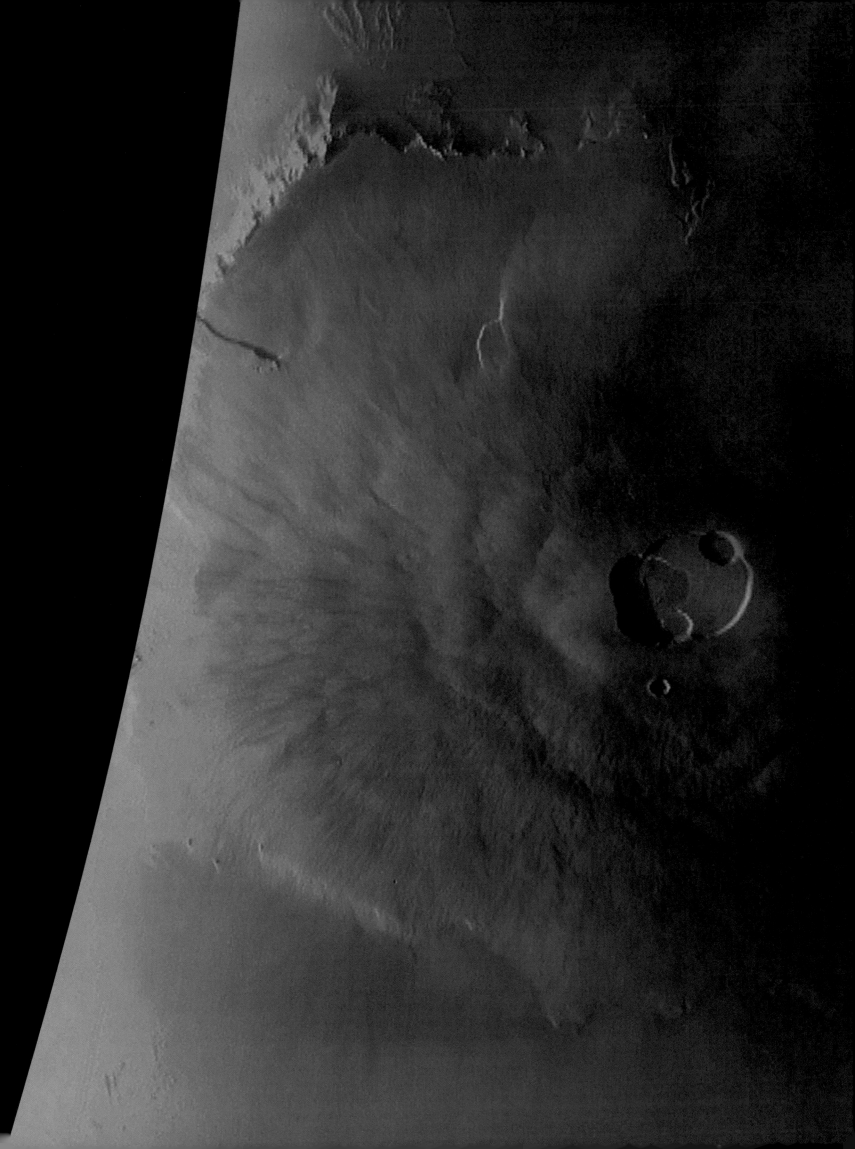

OLYMPUS MONS

Tallest volcano in the solar
system, this Martian Kilimanjaro
rears nearly 17 miles high and
spans almost 400 miles across.
(Earth's Mount Everest rises only
5.5 miles above sea level.) Bluish
haze to the right consists of thin,
high-altitude clouds composed
of water ice crystals.

The image is one of many
from the Mars Global Surveyor.
Launched in November 1996,
MGS went into orbit around the
red planet ten months later.
Almost immediately it began
sending back highly detailed
views and much new information
about forces that may have
shaped the planet's outer layer.
It also detected remnant
magnetic fields in the Martian
crust, possibly indicating that the
planet once had a convecting
and molten core.

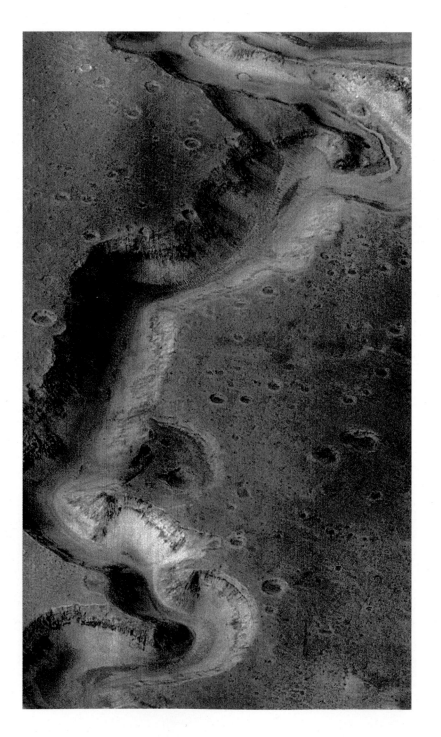

EVIDENCE OF LIQUID WATER?

High-resolution imagery from the Mars Global Surveyor includes a view of Nanedi Valles (left), a canyon about 1.5 miles wide that etches a cratered plateau, revealing layers that seem to have been deposited either by water or by volcanoes. Remarkably, the meanders visible here mirror those of rivers on Earth, including tight curves and even an oxbow bend.

Close-up of a small plateau within Valles Marineris (opposite) shows finely layered rock walls, possibly indicating volcanic or sedimentary origins as well. Such topographic features suggest that liquid water may have flowed on the Martian surface in the past—implying an ancient climate very different from what exists today.

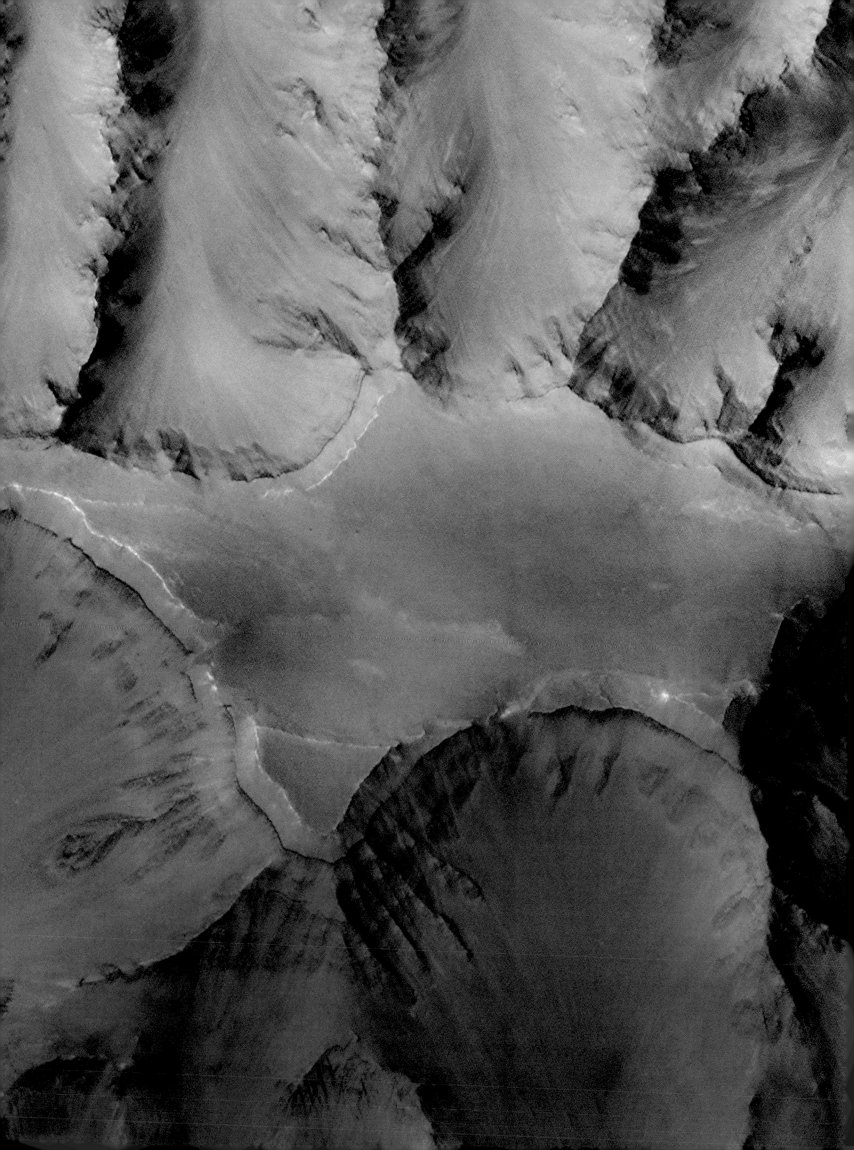

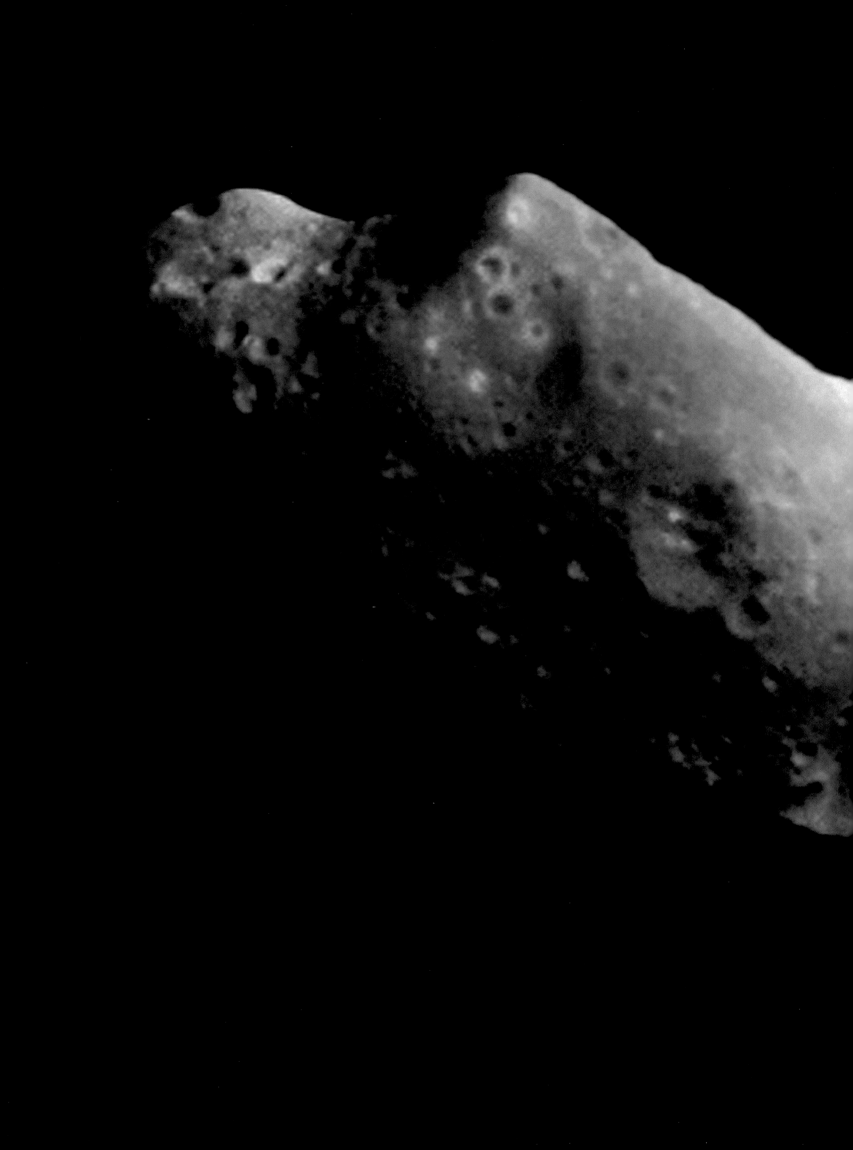

First asteroid found to have its own satellite, potato-shaped Ida boasts a tiny moon named Dactyl, below. Ida measures roughly 35 miles long and 9 miles wide. It, Dactyl, and millions of other rocky residents of the asteroid belt speed around the sun, occupying a broad swath between the orbits of the outermost terrestrial planet, Mars, and the innermost gas giant, Jupiter.

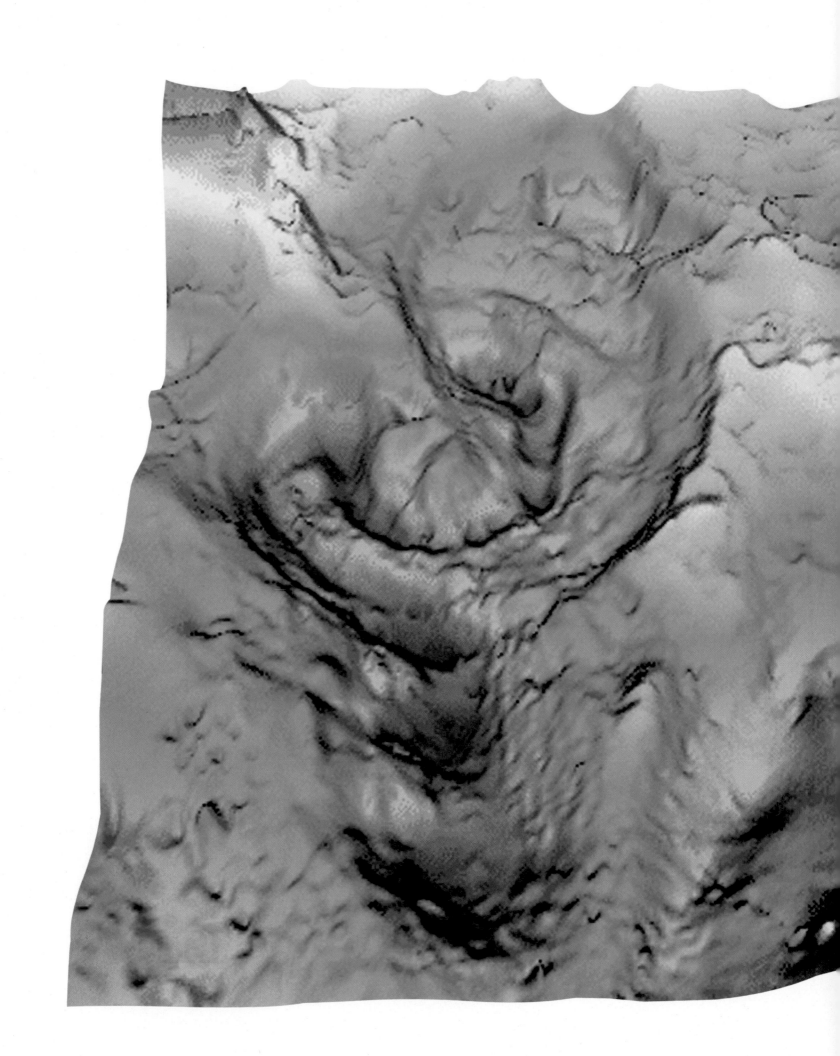

FOOTPRINT OF MASS EXTINCTIONS?

When Earth's Cretaceous Period suddenly ended 65 million years ago, about 70 percent of all species then alive abruptly died out, including the dinosaurs. Scientists now strongly suspect that a colliding asteroid or comet was the culprit. In 1990, an analysis of gravitational and magnetic fields detected the Chicxulub crater (left) deep beneath sedimentary layers in Mexico's Yucatán Peninsula. This 65-million-year-old, 112-mile-wide ring is consistent with scientific expectations for the impact of a 6-to-12-mile-wide asteroid, which would have spawned huge tsunamis, firestorms, and earthquakes, as well as enough debris to cloud the atmosphere for years, drastically altering the short-term environment.

FUTURE NEMESIS?

False-color rendering of a real-life asteroid, this Hubble Telescope image of Vesta (above) shows a rugged surface punctuated by a single crater half as wide as the asteroid itself. Currently some 360 miles across, Vesta may have undergone a tremendous splintering collision about a billion years ago. In October 1960, a small rocky chunk believed to have been part of this asteroid fell to Earth and was recovered in Australia.

REALM OF
THE GIANTS

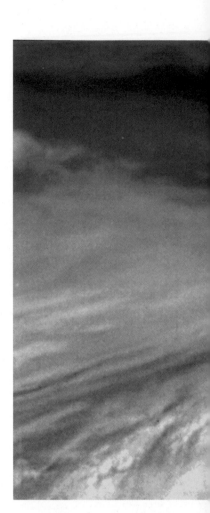

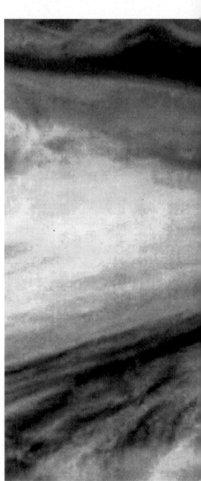

EQUATORIAL HOT SPOT

Chronicling a dramatic "hot spot," or hole in the cloud cover, this simulated true-color view of Jupiter shows a dark blob (right, top). Near-infrared imagery (right, bottom), has been colorized to reveal cloud variations. Blue denotes high, thin clouds; red signifies low ones; white means high and thick. Both images span roughly 21,100 by 6,800 miles.

[PRECEDING PAGES]
Striped by the shadow of its own rings, the gas giant Saturn dwarfs nearby Tethys and Dione, 2 of its 18 known moons.

OUT PAST THE ASTEROID BELT, THE LANDSCAPE CHANGES. Distances get bigger, space gets emptier, the sun gets dimmer. It's about 93 million miles from the sun to the Earth, a distance astronomers call an astronomical unit, or AU. From Earth to Mars is about half an AU, and to the center of the asteroid belt is about one more. But before we get from the belt to Jupiter, we'll have to travel across two more AU, and we won't get to Pluto, the farthest planet, until we've crossed another 34! Compared to this vast emptiness, the space occupied by all four terrestrial planets shrinks into insignificance.

But while the distances between these outer planets are long and empty, the planets themselves make up for it. They are huge. It's not for nothing that astronomers refer to them as gas giants. They're also called the Jovian planets, since Jove is another name for Jupiter, the largest planet of all. You could pack more than 1,400 Earths into it. In addition to their size, these gigantic worlds of our outer solar system travel in company: Each is surrounded by moons—not a paltry one or two, as with the terrestrial planets, but swarms of them, each a unique and interesting world in its own right.

And then there are the rings. The best-known set surrounds Saturn, of course, but all gas giants have them. It's just that Saturn's are composed mostly of water ice, and so they glisten in the wan sunshine of the outer solar system. The rings of the other giants are thinner, darker, and harder to see, even through telescopes.

In the realm of the gas giants, the sun fades to a small, glowing pebble in the sky, a reminder that when the outer part of the solar system was forming, materials like hydrogen and helium weren't driven off as they were on planets nearer the sun. Instead, they managed to linger in huge quantities and were taken up into the bodies of the planets. That's why the giants are so different from the more familiar terrestrial planets.

But enough of theory. Let's get back into our spaceship and begin our tour of the outer reaches of our home system.

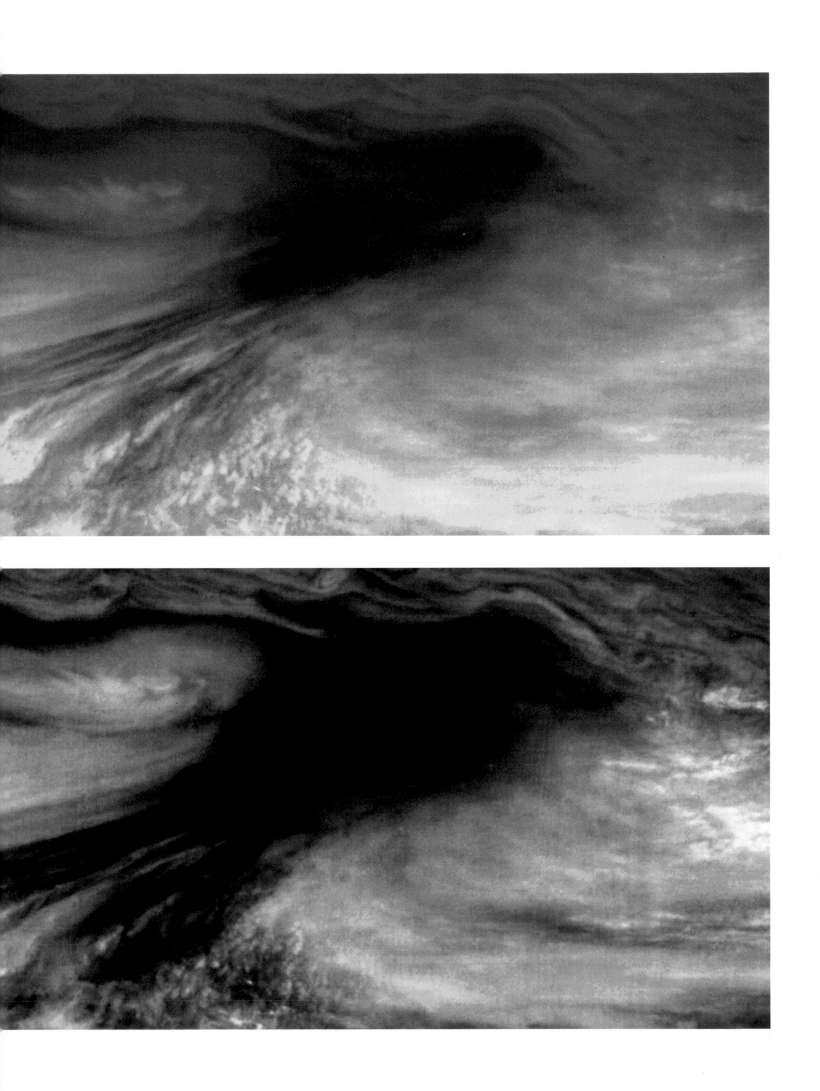

APPROPRIATELY NAMED FOR THE ROMAN KING OF GODS, the largest planet in our solar system—Jupiter—is also the gas giant nearest Earth, and is visible in the night sky even to unaided eyes. It contains about two-and-a-half times the mass of the eight other planets combined, but because its mass is spread over an enormous volume, its average density isn't all that high—a chunk of average Jovian material would barely sink in your bathtub.

Our basic understanding of how Jupiter and the other gas giants formed is that rock-ice cores appeared first, through the process of planetesimal accumulation described in the previous chapter. Then layers of hydrogen and helium were added. Jupiter's rock-ice core—still a purely theoretical hypothesis—is thought to be a few times as massive as the entire Earth. Scientists believe the rest of this planet's interior is differentiated and layered, like the interior of the Earth, but for different reasons. For while Earth formed first as a fairly homogenous mass and then melted and underwent differentiation, Jupiter took shape with its core in place.

Jupiter's enormous mass means it has tremendous internal pressures. The pressure in its core is some 20 times greater than pressures at the Earth's center—and about 70 *million* times greater than the average sea-level air pressure on Earth. Because of this, the substance of this planet—hydrogen, mostly—occurs in unusual forms. On Earth, we normally encounter hydrogen as a gas, but under high pressure, hydrogen atoms can be pushed together to form a liquid or even a metal. Most of Jupiter's interior, in fact, is filled with hydrogen in various liquid and metallic forms. Only in its outer atmosphere do we find hydrogen and helium as gases. Like other gas giants but unlike the Earth, Jupiter doesn't have a hard surface; drop down into it and you pass through layers of cloud and an atmosphere that grows increasingly thicker, until it simply merges into a liquid.

Oddly, Jupiter appears to radiate more energy into space than it receives from the sun—almost 70 percent more. This means it must contain some source of energy. We currently believe that it is still radiating away heat that accumulated back when it formed, since its great size necessitates a longer cooling-off period for heat to leak away gradually. Unlike the Earth, Jupiter derives its internal heat not from radioactivity but from the ancient process of its formation.

Like the other gas giants, Jupiter has a ring system. Its rings are thin, probably no more than 20 miles from top to bottom, and consist of small particles—about the same size as those found in Earth's air on a hazy day. This means that, when viewed in visible light from an Earth-bound telescope, they're essentially transparent. In fact, we didn't know these rings existed until Voyager 1 made its flyby in 1979 and, looking back at the planet, photographed pale rings highlighted by sunlight (see pages 144-145). Nearly 20 years later,

JUPITER

Reddish orange bands grace the largest and most colorful planet in our solar system, which also happens to be the innermost gas giant. Each band indicates a different zone of clouds racing around Jupiter— some eastward, some to the west—at speeds of more than 340 miles per hour. Enormous storms as big as Earth's continents also batter the fifth planet.

measurements taken by the Galileo spacecraft showed that the rings most likely originated when interplanetary debris hit the surfaces of Jupiter's smallest moons, kicking up clouds of dust. These moons are so small—some less than 15 miles across—and their gravity so weak that the tiny bits of rock thrown out in such collisions were lost to them but were subsequently captured by Jupiter and locked into orbit, in effect becoming minuscule moons themselves. Put together enough of these "moons" and you create a ring.

Visually, Jupiter is known for the series of brightly colored bands running across its surface, alternating bright and dark. Lighter areas are places where warm gases from the planet's interior rise, expand, and condense into white clouds; darker bands occur where the cooled "air" sinks back down. These regions are similar to Earth's tropic and temperate zones, and the best way to understand them is to consider for a moment our own planet.

On Earth, more heat from the sun falls at the equator than at the poles. If our planet did not rotate, warm air would simply rise in the equatorial regions, move toward the poles, cool, sink, and move back toward the planet's midsection to complete the cycle. Earth's rotation, however, breaks this cycle into three major cells: the tropics, where surface winds flow east to west; the mid-latitudes, where they go west to east; and the polar regions, where they again flow east to west. The same basic process accounts for Jupiter's many belts and zones. Its great size and rapid rotation—the Jovian day is less than ten hours long—explain why it has half a dozen such regions in each hemisphere, while Earth has only three.

Jupiter's uppermost cloud layers consist of crystals of frozen ammonia, which normally are white. It is believed that relatively small amounts of carbon, nitrogen, sulfur, and phosphorus also occur here, and ultraviolet radiation from the sun causes chemical reactions that create smog, much as it does on Earth. This Jovian smog absorbs varying amounts of violet and blue light, causing regions of the planet to appear yellow and red. Beneath the outer layer of ammonia lies a cloud-free space, then possibly clouds of ammonium hydrosulfide, followed by cloud decks containing liquid water.

Much of what we know about Jupiter's upper atmosphere we owe to a probe launched from the spacecraft Galileo in 1995. Its descent slowed by parachutes, the probe penetrated about 90 miles below the cloud deck, transmitting data for nearly an hour before atmospheric temperatures destroyed it. Because the data had to be sent from the probe to the spacecraft in orbit, and because the probe's entry into Jupiter's atmosphere was difficult, several nervous hours passed before scientists realized that this aspect of Galileo actually worked beautifully. One major surprise: the probe detected less than 30 percent of the amount of water scientists had anticipated. They had

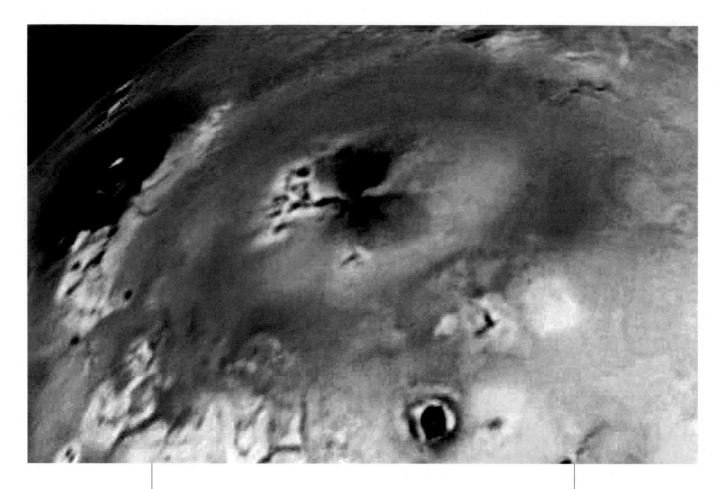

expected that, since Jupiter formed in a frigid region where water was in the form of ice crystals, ice would have been incorporated into the planet when it first took shape. As yet, it is not clear whether the probe happened to plumb a dry spot in the atmosphere—the Jovian equivalent of Earth's Sahara—or whether Jupiter's equatorial region (or perhaps the entire planet) is drier than we thought.

Jupiter's most striking surface feature, the Great Red Spot, is a huge storm in the southern hemisphere. When I say "huge," I mean it— you could drop the entire Earth into the Great Red Spot and barely create a ripple. This storm has raged at least since it was first observed in the 17th century, and probably much before that. Why it started and how long it will last remain unknown.

Jupiter has 16 known moons, ranging in size from spheres thousands of miles across to irregularly shaped lumps less than 20 miles in their smallest dimension. The four biggest and brightest of these—Io, Europa, Ganymede, and Callisto, all mythological loves of the king of gods—were seen by Galileo when he turned his telescope to the sky in 1610; they are called the Galilean moons in his honor. Today, a modern Galileo—the spacecraft, that is— continues to explore the Jovian system.

Each of Jupiter's moons is unique. Formed from clouds of debris and dust by processes similar to those that went on in the solar nebula, these natural satellites are as different from each other as

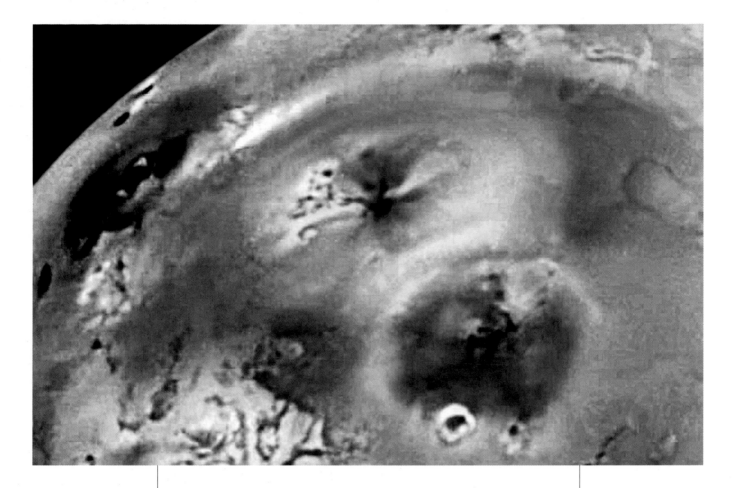

the planets themselves. Each has its own epic of creation to tell. Generally, those closer in are dominated by the gravitational forces of Jupiter, those farther out by the cold of space. Let's look at a few, just to get a sense of the enormous diversity that exists in the mini-solar system that surrounds our largest planet.

THE INNERMOST OF JUPITER'S LARGE MOONS, IO, LOOKS something like a pepperoni pizza (above and opposite): Dark spots punctuate its mottled, yellowish background. They signify active surface volcanoes, and in fact, Io is the only sphere in our solar system—apart from the Earth—known to have them.

On Earth, volcanoes arise when hot magma from the interior finds a way through to the surface. Io is relatively small, however, possessing less that 2 percent of Earth's mass. How can it have enough internal heat to drive volcanoes? The answer lies not in radioactivity (as it does for the Earth), or in leftover heat of formation (as it does for Jupiter), but in the huge gravitational pull of its mother planet. Due to the influence of other Jovian moons, Io's elliptical orbit takes it especially near Jupiter once every 1.77 days. The increased gravitation distorts and flexes Io's body, much like a rubber ball. If you've ever squeezed a rubber ball repeatedly in your hand, you know that flexing generates heat, and it is just this sort of heat that powers Io's volcanism.

BIRTH OF A VOLCANO

Constant change marks the surface of Jovian moon Io, which harbors more than 30 active volcanoes. Images received from the Galileo spacecraft in April (opposite) and October (above) of 1997 monitor the appearance of a large dark spot, evidence of new eruptions amid the reddish surroundings of established volcano Pele. Chemical makeup of the dark material, which blankets an area the size of Arizona, remains unknown.

Voyager first sighted Io's volcanoes during its 1979 flyby, and scientists were astonished at the good fortune of their spacecraft passing Io just as an eruption occurred. Today, we know that there was nothing particularly special about this event, for Io always has a volcano going off somewhere.

Io's volcanoes produce large lava flows and fountainlike plumes of sulfur compounds that produce the moon's orange-and-yellow surface. The darker "pepperoni slices" include both the volcanoes and their lava flows. An Earthling's vision of hell? Maybe, but Io's exotic landscape certainly underscores the diversity of the Jovian system.

NEXT TO IO IS EUROPA. VOYAGER PHOTOGRAPHS SHOWED THIS to be a rather strange world, with an icy surface—real water ice, not carbon dioxide—almost devoid of craters. This seemed to indicate that its surface was rather new. Furthermore, it was crisscrossed with long fissures. Scientists speculated that Europa had frozen over sometime after the great bombardment, effectively erasing evidence of earlier craters, and that the fissures were geological faults.

But the Galileo spacecraft sparked a radical change in that thinking, in the process focusing attention on Europa as one of the solar system's most interesting bodies. As Galileo passed near, at times only 340 miles above Europa's cracked surface, it sent back photographs of unprecedented detail. Instead of an eternally frozen world, these close-ups show something much more like an Arctic ice pack, divided into separate large plates and floes. Furthermore, many of the plates look as though different parts have been broken or rotated, then frozen into place. The long, crisscrossing furrows seem to be places where partially melted ice has risen to the surface, pushed existing ice aside, and then refrozen. This type of history, somewhat reminiscent of the way that rising, molten magma on Earth drives plate tectonics, indicates the possibility that Europa is not totally frozen and could still have a subsurface liquid ocean. From the size of its ice floes, scientists estimate that Europa's icy shell was only a few miles thick when these features formed.

During one pass, Galileo measured Europa's magnetic field. Earth's magnetism, as we've seen, stems from movement in the liquid part of our planet's iron-nickel core. Scientists believe that large-scale magnetic fields can be generated only in this way, and take the absence of such fields as proof that a body is solid. Europa's magnetism might be explained by the presence of a subsurface salty ocean, because salt is a good electrical conductor, enabling the moon to respond to being swept through Jupiter's magnetic field by producing a magnetic field of its own. It is this induced field that the spacecraft detected, and in fact there is some preliminary evidence for the same kind of magnetic effect on the Jovian moon Callisto.

For me, at least, the most convincing indication that an ocean exists on Europa came in 1998, when scientists announced the result of yet another close-up photo; this contained an image of a 20-mile-wide crater named Pwyll, after a character in Welsh folklore. Because it was fairly recent, with an estimated age of only ten million years, it should have retained its original deep, bowl-like shape, for a dead world would have no weathering or erosion following an impact. Yet close examination of Pwyll showed that its floor was smooth and the same level as the surrounding area, with only the ring of crater walls sticking up to mark its presence. This was highly unusual—on our moon, a crater the size of Pwyll would be as deep as the Grand Canyon. The only reasonable explanation for such a crater is that slush from the interior of Europa flowed into it soon after it formed, filling it and then freezing solid.

So our picture of Europa has changed dramatically in the 1990s. Its interior, like Io's, has been warmed despite its great distance from the sun, by interacting with Jupiter's gravitational field. Which means that, instead of being a cold, dead world circling a distant planet, it could be the only other place in the solar system where liquid water currently exists, and possibly has existed for long periods of time. This, of course, brings up the question of extraterrestrial life.

In the last chapter we saw that life on Earth almost certainly arose in the oceans, and that any hope of finding fossils on Mars depends critically on surface water being present there sometime in the past. Given that Earthly life arose very quickly once conditions were right, it is reasonable to ask whether, during the billions of years that Europa's ocean could have existed, living things appeared there as well. Europan life wouldn't have to be complex—even a single-celled bit of pond scum would be a major scientific discovery. And so, our focus has changed; we no longer see Europa as just another one of Jupiter's 16 known moons, but as a possible home for life. In 2003, NASA plans a mission that will orbit Europa in hopes of producing a detailed map of its surface, much like the one we now have for Mars. Some NASA engineers are even looking forward to the possibility of a future probe that would land on the Europan surface, drill through its miles of surface ice, and confirm the presence of an ocean. The ultimate hope would be to bring a sample of that ocean to Earth for study.

GANYMEDE, THE THIRD LARGE MOON OUT FROM JUPITER, also has been cast in new light by the Galileo mission. Named for the cupbearer to the gods, Ganymede is the solar system's largest moon, measuring more than 3,200 miles across. Surfaced largely with water ice, its terrain consists largely of craters and grooves, with no hills more than half a mile high. Judged by its relative flatness, Ganymede would be easy to interpret as a frozen, inactive world.

But in 1996, as the Galileo spacecraft encountered Ganymede, our picture of this moon became much clearer. The grooved regions, upon higher magnification, showed that Ganymede's crust was relatively thin; tearing and deformation apparently had created the grooves. Even more importantly, Galileo was able to detect and measure an internally generated magnetic field around Ganymede. This moon may have a layered structure with a molten iron core, making it in essence a smaller version of our own Earth. One current theory suggests that well after its formation, Ganymede underwent orbital changes that intensified Jupiter's gravitational pull and caused the moon to flex and heat up. Conceivably, sufficient warming could have driven the process of differentiation, producing what today seems a moon with an icy exterior but a molten center.

In addition to Callisto, at least 12 other interesting moons circle Jupiter, but it's time to head on to the next planet out—the one astronomers have called "the jewel of the solar system."

IN 1610, GALILEO TURNED HIS TELESCOPE TOWARD THIS, THE most distant planet visible to the naked eye, and became one of the first humans to see what many still consider the most beautiful sight in our solar system. Galileo's primitive telescope didn't offer a very clear view; to him the rings looked more like fuzzy ears sticking out from the main body. He even thought he might be looking at a triple planet. Only years later was it established that the "ears" were really a ring encircling the planet.

Like Jupiter, Saturn makes a bright, starlike appearance in the night sky, though its colors are paler, thanks to a hazy outer layer. It, too, is a gas giant, with much the same structure and composition as its somewhat larger neighbor—that is, a rocky or icy core surrounded by layers of compressed metallic and liquid hydrogen, topped with an atmosphere that is primarily hydrogen and helium. It has less than a third of Jupiter's mass, however, so it is much less dense. In fact, it is less dense than water; if you could find a bucket big enough, Saturn would actually float.

Saturn also rotates rapidly—about once every ten hours, which gives its atmosphere a belt-and-zone structure similar to Jupiter's, and its outer cloud layer consists of frozen crystals of ammonia. But because it is farther from the sun and thus colder than Jupiter, its atmosphere is overlaid with a high-altitude haze that mutes the colors of its colorful belts and zones. Beneath these ammonia clouds lies a clear layer, floored by a deck of ammonium hydrosulfide clouds. Below that, presumably, are clouds of ordinary water ice.

Another similarity between these two largest planets is that both radiate about twice as much energy into space as they receive from the sun. Some scientists argue that Saturn, like Jupiter, is still getting rid of heat accumulated when the planet first formed. Others feel that liquid drops of helium may be forming in its interior, creating a kind of rain that releases heat in the process. If this is truly the case, then Saturn is still undergoing differentiation.

SATURN'S STRIKING AND EVOCATIVE TRADEMARK, OF COURSE, is the set of rings that circle its equatorial plane. First recognized as such by Dutch astronomer Christiaan Huygens in 1659, they have been the object of avid study ever since. In 1675, astronomer Jean - Dominique Cassini noted that what at first had seemed a single broad ring actually was two narrower but concentric ones separated by a gap, which now bears his name. We now know that smaller gaps and many other features also subdivide the rings, some of which are the result of the gravitational effects of Saturn's innermost moons, orbiting just beyond the rings.

Because the rings (and Saturn itself) are tilted relative to the planet's orbital plane, their angle seems to vary as Saturn moves around the sun. Viewed edge on, they are so thin that they almost disappear—some have been estimated to be only 60 feet thick!

By the end of the 19th century, astronomers had realized that the rings were neither a continuous solid nor a continuous liquid. Today, we know they are made up of many individually orbiting clumps of dirty ice, ranging in size from microscopic snow to house-size boulders. For all their spectacular appearance, their total mass is about three-thousandths that of the moon. With so little mass spread over such a huge region of space, they're really not very substantial. Nevertheless, because ice reflects light so well, the rings announce their presence with panache.

SATURN HAS AT LEAST EIGHTEEN MOONS, SEVERAL OF THEM discovered by the two Voyager spacecraft. Like Jupiter's satellites, each is a unique world, with its own history and distinguishing characteristics. By far the most interesting is Titan, the 15th out from the mother planet, first seen by Christiaan Huygens in 1655.

Titan is bigger than Mercury, but it possesses only 40 percent of that planet's mass, and it appears to be about half ice, half rock. What truly sets it apart from all other moons of the solar system, however, is that it actually has an appreciable atmosphere, and a rather dense one at that. This Titanic atmosphere is mostly nitrogen, along with some methane—a major component of natural gas on Earth—and other organic compounds that have condensed and polymerized, giving this moon the reddish orange color so apparent in Voyager photographs. Think of Titan's atmosphere as a kind of heavy smog, so thick that it exerts surface pressures that are half again as high as sea-level pressures on Earth.

SATURN

Least dense of all the planets, this grand lord of the rings has a specific gravity of only 0.7, making it lighter than water. Like Jupiter, it is mostly hydrogen and helium, with traces of water, methane, and ammonia, all shrouding a rocky core. Its many rings consist largely of "dirty ice"—frozen water and dust.

We now believe that Titan's atmosphere produces a greenhouse effect, warming the moon's surface to minus 180°C—only about 90° above absolute zero, but balmy relative to outer space. There is no possibility of liquid water at this temperature, of course, but scientists have suggested that oceans of liquid methane or ethane might exist on Titan's surface. Both substances are highly flammable gases at Earthly temperatures and pressures, but with Titan's high pressure and low temperature, they should be liquids there. One reason scientists believe such oceans could exist is sunlight. Even the weak sunlight that reaches Titan can break down methane, and the fact that Titan continues to show appreciable methane in its atmosphere implies that its supply is constantly being renewed. What better reservoir than a hydrocarbon ocean beneath its smoggy clouds?

The prospect of liquid oceans combined with an atmosphere rich in organic molecules conjures images of the early Earth, when organic materials probably rained into the oceans, providing a broth that eventually led to life. At Earthly temperatures, the chemical reactions required to produce living systems take place fairly rapidly. But on frigid Titan, they could proceed only slowly—if at all. That's why we feel that, if we could see what was happening on Titan's surface now, we would learn something about those early days on our own planet.

So there's an important difference between Titan and Europa. Even though both are far from the warming sun, even though both have surface temperatures well below freezing, Europa, as we've seen, has an internal heating mechanism. This means that if Europa has an ocean, in places its temperature actually could be above freezing. This isn't very different from temperatures on the early Earth, so scientists argue that it is possible that the same chemical reactions that led to life on Earth could have taken place there. Europa, in other words, is a place where life might exist right now.

Titan, even a Titan with oceans, does not have this possibility simply because it is too cold. Chemical reactions that take a few thousand years in liquid water might take millions or even billions of years in frigid liquid methane. Consequently, most scientists hope to find on Titan not life itself but the chemical reactions that could lead to life, running in slow motion because of the cold.

In 1990, astronomers bounced radar beams off Titan's surface from transmitters on Earth and analyzed the return waves. They found no evidence of methane or ethane oceans covering the entire moon, but smaller lakes and ponds weren't ruled out. Then the Hubble Space Telescope, from its orbital perch high above Earth's distorting atmosphere, looked at Titan in infrared light. The results: a surface of alternating light and dark patches. One interpretation is that those darker areas may be hydrocarbon lakes, while the lighter regions are continents of frozen water and ammonia.

Saturn's moons and even its rings aren't as well understood as Jupiter's because, as yet, there's been no space mission dedicated solely to its study. But that situation is changing. As I write this, the Cassini spacecraft is speeding toward a rendezvous with Saturn in 2004. Launched in 1996, Cassini follows a complex trajectory that involves "slingshot" flybys of Venus (twice), Earth, and Jupiter before arriving at its ultimate goal. Each planetary encounter boosts Cassini on its way, thereby reducing the amount of fuel that has to be carried.

Cassini is expected to do for Saturn what Galileo did for Jupiter. It also carries a probe, named Huygens, that will be dropped into the atmosphere of Titan and descend by parachute, giving us our first look at that atmosphere and—with luck—our first look at oceans and shorelines that don't involve water. In addition, Cassini will make detailed measurements of Saturn's magnetic field, outer atmosphere, rings, and icy moons during its four-year sojourn around this spectacular gas giant.

ONCE WE CLEAR THE ORBIT OF SATURN, WE'RE BEYOND THE range of the unaided human eye, out among planets that can be seen only with a telescope. We're also in a realm associated with one of the most exclusive clubs in history—those human beings who have discovered previously unknown planets in our solar system. It's a club with only three members.

Uranus was first sighted in 1781, more or less by accident, by William Herschel, a German-born military musician and church music director turned amateur astronomer. While scanning the sky from his backyard in the city of Bath, England, Herschel noticed an unusual object that seemed to be more a disk than a point of light, more like a planet than a star. Later measurements showed that it moved just the right amount each night to be a planet beyond Saturn. Herschel proposed that the new planet be named Georgium Sidus, or George's Star, in hopes of getting support from King George III. Although the astronomical community eventually settled on the name Uranus, after one of the primeval Greek gods, Herschel's ploy worked perfectly—he got a life pension from the Crown that enabled him to become a full-time astronomer and telescope maker.

In quite a few ways, Uranus is an oddball. For one thing, its axis of rotation is in the plane of its orbit—in essence, it spins on its side. This means that for half of the 84 years it takes Uranus to make a complete orbit of our sun, its north pole is in daylight; for the other half, it's in darkness. This planet has its own magnetic field but, for reasons we do not yet understand, the poles of that field seem to be off-center. They're also tilted 59 degrees with respect to the axis of rotation. That's the equivalent, on Earth, of having our magnetic north pole relocated to Florida.

URANUS

Third in size after Jupiter and Saturn, bluish-green gas giant Uranus also consists largely of hydrogen and helium, and boasts numerous moons as well as a ring system—although its axis of rotation is severely tilted relative to the rest of the solar system. Atmospheric methane and a high-altitude, photochemical smog contribute to its color.

Uranus also is the smallest of the gas giants, with less than one-twentieth of Jupiter's mass, and it has a different internal structure from its big brothers. For although it has a rocky core, its mantle consists of icy materials like water, methane, and ammonia, perhaps mixed with rock. Its atmosphere is mainly hydrogen, helium, and methane, and it's cold—minus 214°C. From Earth, it appears bluish-green and rather featureless, wrapped in cloud. Its color stems from a methane-rich atmosphere, which absorbs reds from the weak sunlight that reaches it, reflecting back blues and greens. Its seemingly indistinct surface is due to an outer deck of ammonia ice clouds overlaid by a thick haze composed primarily of hydrocarbon ices.

Five moons were known to be orbiting Uranus before the Voyager flyby in 1986; ten more were found by Voyager, and another two have been discovered since then. Astronomers have named a number of them after Shakespearean characters, from Titania and Oberon—each about a thousand miles in diameter—to Puck, a mere hundred miles across, and even smaller satellites.

Equally interesting is the Uranian ring system. In 1977, astronomers were observing a star fade as the leading edge of Uranus's atmosphere passed in front of it. By watching how that star's light was absorbed, they hoped to learn the temperature of the Uranian atmosphere.

Some of the scientists were observing Uranus through a telescope mounted in an airplane, because they wanted to get above the clouds in Earth's atmosphere, which could throw off their results. Much to their surprise, they saw the star dim and then brighten several times before it aligned with the atmosphere of Uranus. The inescapable explanation: Starlight was being blocked by previously unknown rings of Uranus. No one had seen these rings because they are fairly narrow—the biggest is only 60 miles wide—and very dark, reflecting only a few percent of the very weak sunlight that hits them.

Thanks to Voyager photographs, we now know that Uranus has at least ten of these dark rings. Most of the rings are merely a few yards in thickness. They are thought to be composed mainly of methane ice coated with carbon, which explains why they're so black.

MOVING STILL FARTHER OUT, WE COME TO NEPTUNE, THE LAST gas giant as well as the last planet visited by Voyager 2. As with Uranus, the story of its discovery is as interesting as the planet itself. Once Uranus had been found, astronomers attempted to plot its orbit. Try as they might, however, they couldn't get it to move the way it should, according to Newton's Law of Universal Gravitation. By the mid-1800s, this problem had grown into a major scientific challenge. Newton had taught us to think of the universe as a great clock whose gears followed predictable rules, yet here was a "gear" that resolutely refused to fit in.

NEPTUNE

Cold, distant, and blue, this outermost gas giant is also the densest, indicating a relatively large core. Its color stems from a methane-rich atmosphere.

Radiating far more energy into space than it receives from the sun, Neptune is home to titanic storms and winds of more than 1,000 miles per hour—among the strongest in the solar system.

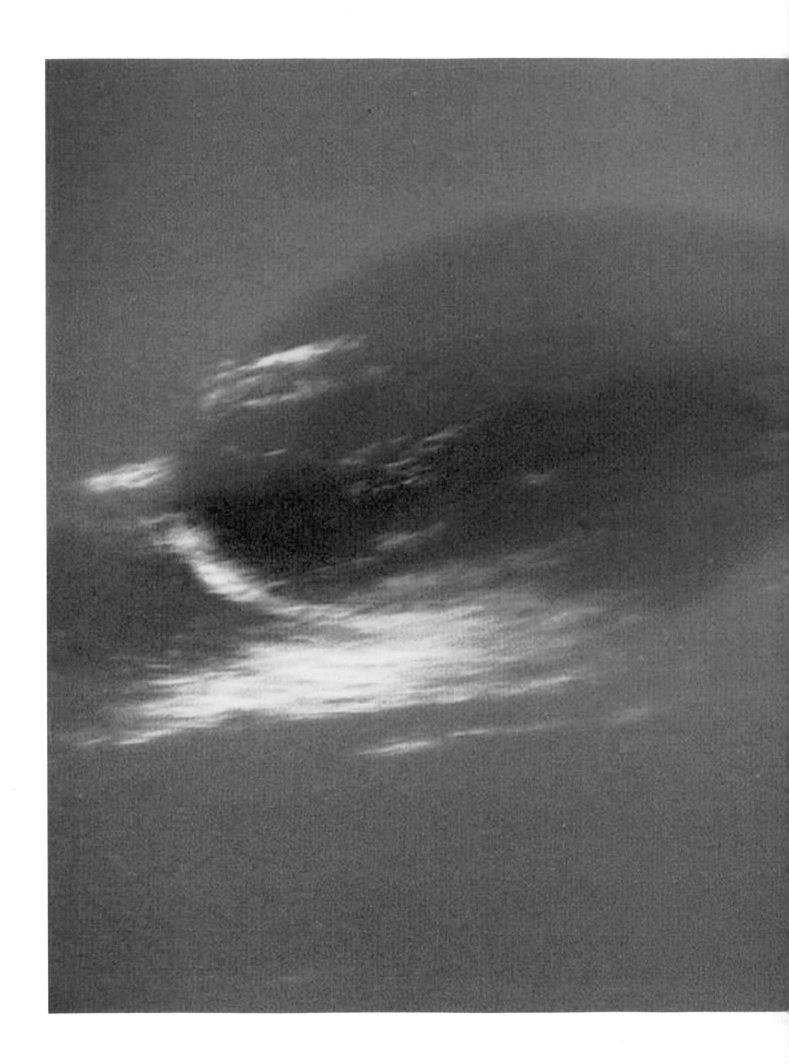

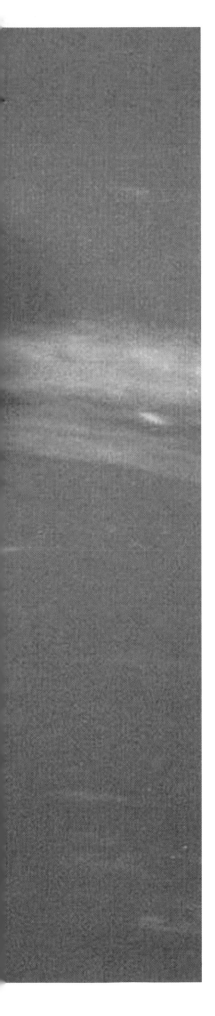

A way out of this dilemma appeared simultaneously to two scientists—John Couch Adams in England and Urbain Leverrier in France. Both realized that the deviations in the orbit of Uranus would make sense if there was another planet beyond Uranus, pulling it away from its expected Newtonian track. Each man carried out the calculations that predicted just where this unknown planet had to be—in effect, they told astronomers where to point their telescopes if they wanted to make an important discovery. Both were initially ignored. Finally, in 1846, Johann Galle in Berlin turned his telescope to the area predicted by their calculations—and became the first person to see the planet we now call Neptune.

In the history of science, this was as stunning an event as sighting the return of Halley's comet. For it showed conclusively that the magnificent Newtonian picture of the cosmos, with worlds moving in stately orbits according to well-defined laws of gravity, could be used to predict things that had never been seen.

Slightly larger than Uranus and significantly smaller than Jupiter and Saturn, Neptune seems to be a middling kind of gas giant. Yet it radiates almost three times as much energy into space as it receives from the sun, which means that, like Jupiter and Saturn, it must have some internal source of heat, perhaps heat left over from its formation. Thus it is at about the same temperature as Uranus, even though it is farther from the sun. Like Uranus, its magnetic field—which is about as strong as Earth's—is tilted significantly from its axis of rotation. No one knows why.

Named after the Roman god of the oceans, Neptune has a bluish cast due to the methane in its atmosphere, which consists mostly of hydrogen and helium. Voyager 2 flew by this planet in 1989, giving scientists close-up views of the upper atmosphere. They found high, streaky, cirrus-like clouds poised about 45 miles above a main cloud deck of hydrogen sulfide and believe that those high clouds are composed of methane, frozen into a solid form and lofted up by currents driven by the planet's internal heat.

Voyager also witnessed a great storm on Neptune reminiscent of Jupiter's Great Red Spot. Later dubbed the Great Dark Spot, this storm was about 6,000 miles across and was located in the planet's southern hemisphere. We still don't understand what caused the Red Spot and Dark Spot. But we have learned that winds on Neptune are among the strongest in the solar system, with speeds in some cases exceeding 1,000 miles per hour. (For comparison, the fastest winds detected on Earth—at Mount Washington, New Hampshire in 1934—stand at 231 miles per hour.)

Voyager discovered six new moons orbiting Neptune, apart from two others already known. It also confirmed the existence of a ring system around the planet. Up to that time, Earth-based astronomers

MYSTERIOUS DARK SPOT

Discovered in 1989 by Voyager 2, Neptune's Great Dark Spot seems to be a system similar to Jupiter's Great Red Spot but far more variable. In fact, when viewed five years later, the spot was nowhere to be found, while another storm then raged in the planet's northern hemisphere. Voyager also detected surprising turbulence on Neptune, in addition to zonal cloud bands and pulses of radio emissions.

had observed objects around Neptune they dubbed "ring arcs." Voyager showed that most of these objects were denser segments of a single, diffuse ring. Like all gas giants, Neptune has several rings; the three most prominent have been named for the trio that helped find this planet: Adams, Leverrier, and Galle.

ANY WAY YOU LOOK AT IT, PLUTO IS WEIRD. IT IS A SMALL, rocky orb out beyond the realm of the gas giants. While most planets in our solar system orbit the sun in the same plane (called the ecliptic), Pluto's orbit is tilted at an angle of 17 degrees to that plane. Also, its orbit is so eccentric that, between 1979 and 1999, it actually moved *inside* Neptune's orbit—so, for that brief period, it wasn't even the farthest planet from the sun.

Pluto was discovered in 1930 by a young and avid amateur astronomer named Clyde Tombaugh. Growing up on a farm, Tombaugh was the sort who would build his own telescope. He sent some sketches he'd made of Mars to the Lowell Observatory in Arizona, more or less to ask how he was doing. The drawings were so good that he was quickly offered a job at Lowell, searching for what was then called Planet X. Astronomers had detected some deviations from predicted values for Neptune's orbit and thought they might be due to another, yet-undiscovered planet beyond. In fact, we now know that their measurements were incorrect, and that they had no evidence in the late 1920s for such "deviations." Nevertheless, Tombaugh began a painstaking process of photographing one area of the sky over a period of many days and looking for an object that moved by just the right amount to be a planet. On February 18, 1930 he hit pay dirt, discovering the planet that we now call Pluto, in honor of the Roman god of the underworld.

In 1978, astronomer James Christy of the Naval Observatory in Washington, D.C., methodically measured the position of Pluto on photographic images and discovered that it had a moon. The moon was named Charon, after the mythological ferryman who transported dead souls across the river Styx. Because Charon's speed depends on the mass of Pluto, we can make a fairly good determination of that mass by measuring the time it takes Charon to complete a full orbit. Pluto turns out to be only 1/500 as heavy as Earth, and smaller than Mercury in size, making it far too little to produce the once-suspected changes in Neptune's orbit. In fact, in density and size both Pluto and Charon are very similar to the moons of Neptune. As we shall see, this has led some scientists to propose that Pluto and Charon were not formed by the same processes that created the other planets and moons in our solar system.

At the moment, Pluto is relatively near the sun. Its atmosphere is very thin, composed mostly of nitrogen, while a layer of methane,

PLUTO

As mysterious and daunting as the Roman god of the underworld for whom it was named, the solar system's tiniest and most remote planet was discovered by Clyde Tombaugh in 1930. Unlike the gas giants, it is small— smaller than Earth's moon—and rocky. Current speculation holds that Pluto may have originated in the disk-shaped Kuiper belt, beyond the gas giants.

nitrogen, and carbon monoxide ices coats its surface. Scientists expect that as Pluto moves away from the sun, the methane in its atmosphere will freeze and fall as a kind of snow. When this happens, the planet's surface will brighten considerably, since liquid methane reflects light almost totally. Some astronomers even claim that, if they could sit on Pluto's surface, they would be able to read a newspaper in all the light reflected by the methane snow. In contrast, Charon's surface appears to be mostly water ice.

Despite recent photographs by the Hubble Telescope, Pluto remains one of the least studied and least understood objects in our solar system. It has never had a flyby, but in 2004 NASA will launch the Pluto-Kuiper Express, which will map the planet and its moon in detail. There have been many obstacles to such a mission, not the least of which is time: It will take about a decade for the spacecraft to reach this frozen planet. Consequently, we may be well into the next century before we know substantially more about Pluto and its companion than we can learn from the Hubble Telescope.

OUT PAST THE ORBIT OF NEPTUNE WE FIND A REALM WHOSE very existence was in doubt until recently. This is the so-called Kuiper belt, named for Dutch-American astronomer Gerard Kuiper, who suggested it in 1951. (Kenneth Edgeworth, an Irish scientist, put forth the same idea even earlier, in the 1940s.) The Kuiper belt turns out to be a flat ring composed of big but thinly scattered lumps of ice mixed with dust and other materials. Think of them as big and dirty snowballs, dozens or even hundreds of miles across. They serve as the nuclei of comets, and occasionally one of them breaks loose from the belt and ranges closer in, to where we can see it.

Kuiper proposed that comets could have formed in this distant, doughnut-shaped area of our solar system. Most astronomers of his day, however, believed that all comets came from another reservoir even farther out in space known as the Oort cloud, which we will visit in a moment. In 1980, Julio Fernandez proposed that the belt was a direct source of short-period comets, those which orbit the sun in 200 years or less. It wasn't until 1992 that two astronomers, working at the University of Hawaii telescope on Mauna Kea, actually sighted an object in the Kuiper belt. The astronomical community officially named it 1992 QB_1, though its discoverers wanted to call it "Smiley," after John LeCarré's fictional spy who came in from the cold.

Since then, more than 130 objects have been found, all ranging from 50 to 300 miles across. Based on this sampling, it is estimated that the Kuiper belt contains at least 35,000 "snowballs" of similar size. If that's true, then it contains much more matter than was first thought, hundreds of times more than the better-known asteroid belt between Mars and Jupiter.

Astronomers now believe that at one time the Kuiper belt was even more extensive, reaching as far in as the orbit of Neptune. Neptune's gravitational tug, they argue, systematically pulled those "near" comets out of the belt over the last 4.5 billion years, and nudged them into the inner solar system. Most such comets simply evaporate away in 10,000 years or so, but as we saw with Shoemaker-Levy 9, they also can collide spectacularly with a planet or even the sun. The gradual stripping away of material from the inner side of the Kuiper belt is called gravitational erosion. One very intriguing supposition to come from this dynamic picture of our outer solar system is that Pluto and its moon, Charon—and possibly even Neptune's moon, Triton—have an origin different from the solar system's other planets and moons.

It may be, say proponents, that Pluto is a surviving remnant from the original Kuiper belt, orbiting with similar but perhaps smaller objects in the belt over eons, as gravitational erosion left it in lonely splendor. In this scenario, Pluto captured Charon early in its lifetime. Similarly, Triton might be a former Kuiper belt resident that long ago was captured by the gravitational pull of Neptune.

This would explain many of Pluto's peculiarities. But we must realize that the Kuiper belt is the source of only some comets, not all. Comets enter our solar system from all directions, not just from the flat, ecliptic plane in which the Kuiper lies. To understand the rest of the comet story, we need to go still farther out into the solar system, to the Oort cloud. Before we do, however, we should note that the outer solar system harbors all sorts of interesting things, including a few that are absolutely unique. They aren't very big—a few yards across, weighing less than an old Volkswagen Beetle. Like the Beetle, they are man-made—the most distant human artifacts in the universe. I'm referring to the Pioneer and Voyager spacecraft. Launched in the 1970s, these diminutive explorers gave us our first look at the outer reaches of our solar system. Today, way beyond the orbit of Pluto, they are plunging ever farther into space, each day going where no man—or man-made object—has gone before.

Voyager 1 is now the farthest out, moving away from the sun at about 300 million miles per year. Scientists expect that, around the year 2010, it will reach the heliopause—a magnetic no-man's-land where the sun's solar wind and magnetic field equal the magnetic field of the galaxy itself. While the sun's gravitational pull extends far beyond this zone, the heliopause is one defining "edge" of the solar system, roughly 100 AU from the sun. Once there, Voyager will be well on its way to leaving its home system behind.

About a decade after that, sometime around 2020, the power remaining in Voyager's generators will drop below levels needed to run all of its instruments, and some instruments will be turned off. At this time, Voyager 1 should be 138 AU from the sun, while

Voyager 2 will be 113 AU out. They will just keep on going, drifting farther and farther into the cold and dark of interstellar space.

But I don't think we've seen the last of them. I think that a few hundred years from now a starship will pull up next to the Voyagers, take them aboard, and bring them home. Then, after appropriate scientific study, they will be put on display at the Smithsonian, a fitting monument to the beginning of the era of space exploration.

EVEN AFTER LEAVING THE KUIPER BELT, WE ARE NOT YET IN interstellar space, just in the boondocks of our solar system, a cosmic version of frozen tundra, where the sun shines no more brightly than does Venus in Earth's evening sky and the average temperature is near absolute zero. This is the Oort cloud, named after another Dutch astronomer, Jan Oort, who first predicted its existence in the 1950s. Unlike the disk-shaped Kuiper, the Oort is a spherical shell surrounding the solar system and extending many billions of miles, from Kuiper's outer reaches halfway to the nearest star. It is a vast reservoir of dirty snowballs, some of which may someday appear as comets in our inner solar system. Astronomers estimate that there are some six trillion comets out here, each several miles across, perhaps amounting to a total mass about 40 times that of the Earth. The Oort cloud is so unimaginably huge, however, that each snowball is typically tens of thousands of miles from the next.

Jan Oort deduced the presence of this cometary reservoir by studying long-period comets—those with periods longer than 200 years. He traced their orbits back in time, taking account of the gravitational pull of the outer planets, and found that all of them, though they entered the inner solar system from every direction, seemed to have originated far beyond Pluto's orbit.

Today, with the help of modern computers and much more data, astronomers have determined that these comets originate about 40,000 AU out. The average long-period comet from the Oort cloud has an orbit that brings it through our solar system once every few million years or so. It is this sort of calculation that gives us confidence that the Oort cloud really exists, because its comets are much too far away and too small to be seen by telescope until they enter the inner solar system.

In a sense, the discovery of the Oort cloud is a story that begins with the ancients. For most of human history, comets were seen as strange and unpredictable things. They often were considered omens, good or bad. For example, in 1066, the year of the Norman invasion of England, a comet that we now know was Halley's appeared in the sky. It proved to be an excellent omen for William the Conqueror— but it wasn't very good for Harold the Saxon. As always, the human interpretation of heavenly events is open to a lot of variation.

THE ALLURE
OF COMETS

Peripatetic residents of the vast Oort cloud and other realms, comets have fascinated people for millennia, often inspiring interpretation as omens.

Perhaps the most famous of these repeat visitors, Comet Halley, returns to the inner solar system every 76 years. Photographed from California's Mount Wilson Observatory in May 1910 (right), it flaunts a distinct head and well-formed tail.

Why do comets in the Oort cloud, which is so widely dispersed and so far from us, ever come near Earth? Why don't they simply spend eternity wandering in the cold emptiness of the solar system's farthest reaches, locked into their orbits by the distant but ever-present pull of the sun? In fact, this is exactly what would happen if there weren't outside influences on the Oort cloud. But many things in our galaxy can jostle Oort cloud comets just enough to send them sunward. The most common of these is both obvious and unexpected: Every once in a while, a star passes near enough to our solar system to exert significant force on comets in the cloud. It's been estimated that perhaps a dozen stars pass within a few thousand AU of our sun every million years. Comets that enter our solar system, of course, are affected by the gravitational pull of the planets as well as of the sun. Some are thrown out of the solar system entirely, while others enter orbits that bring them nearer the sun at periodic intervals. Whenever we see a comet in the sky, we should take it as a reminder that there are forces that go far beyond our own solar system and reach to the stars themselves. Jan Oort offered a picturesque description of his namesake cloud: "a garden, gently raked by stellar perturbations."

Our solar system will probably experience just such a perturbation in about 1.4 million years, when a star known as Gleise 710 is expected to pass through the outer parts of the Oort cloud and come within 70,000 AU of the sun. Current estimates put the number of comets due to Gleise 710 at only about 50 percent above normal.

Astronomers believe, however, that there have been times when major perturbations have increased the rate of comets entering the solar system by several hundred percent. Such events, called comet showers, can have dramatic consequences. Some scientists even suggest that the impact that wiped out the dinosaurs was generated not by an asteroid, but by a large comet that was part of a major comet shower 65 million years ago.

Even if that was the case, a nagging question remains: Where did all those Oort cloud comets come from? They couldn't have formed in the orbits they now occupy, because the nebula that gave rise to our solar system should have been much too diffuse in its outer regions for so many objects to have been created. One current theory holds that comets now in the Oort actually formed in the inner solar system and were then thrown into their present position by various gravitational interactions with the planets. Depending on which theorist you talk to, comets could have formed anywhere from the asteroid belt to those lonely stretches past the orbit of Neptune.

As we pass the outermost comets circling in the cold and the dark of this mysterious cloud, we at last truly take leave of our solar system. Ahead lies the vastness of the Milky Way and its billions of stars; beyond that lie all the other galaxies of the cosmos. ●

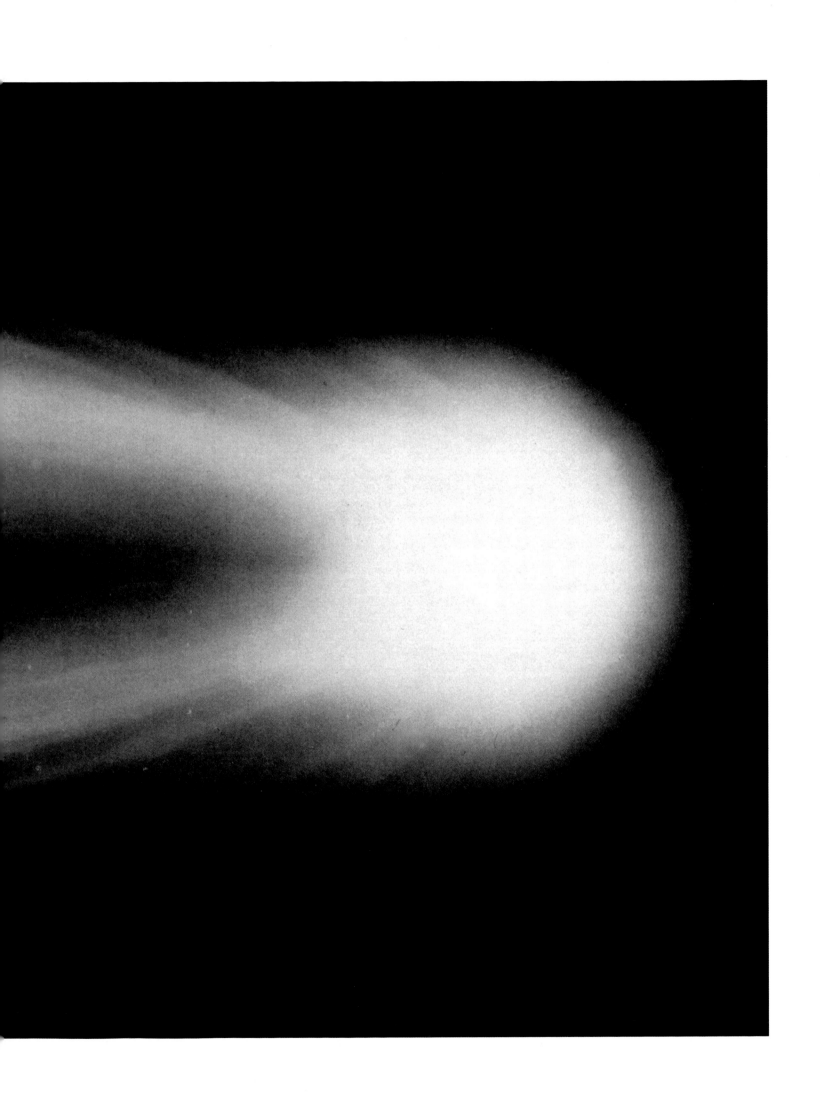

JUPITER

Colorful enigma, Jupiter has fascinated astronomers for centuries
with its ever-changing cloud banks. Composed primarily of hydrogen
and helium, and averaging minus 125°C in temperature, the clouds
should be colorless. Current theory holds that their hues are due to
the action of the sun's ultraviolet light on trace compounds, generating
colored smog particles that grow and deposit on underlying clouds.

Jupiter's darker bands are called belts; the lighter ones are zones.
Winds blow eastward along the equatorial side of each band
and westward along its polar edge, providing the driving force for
tremendous storms such as the Great Red Spot, visible here.

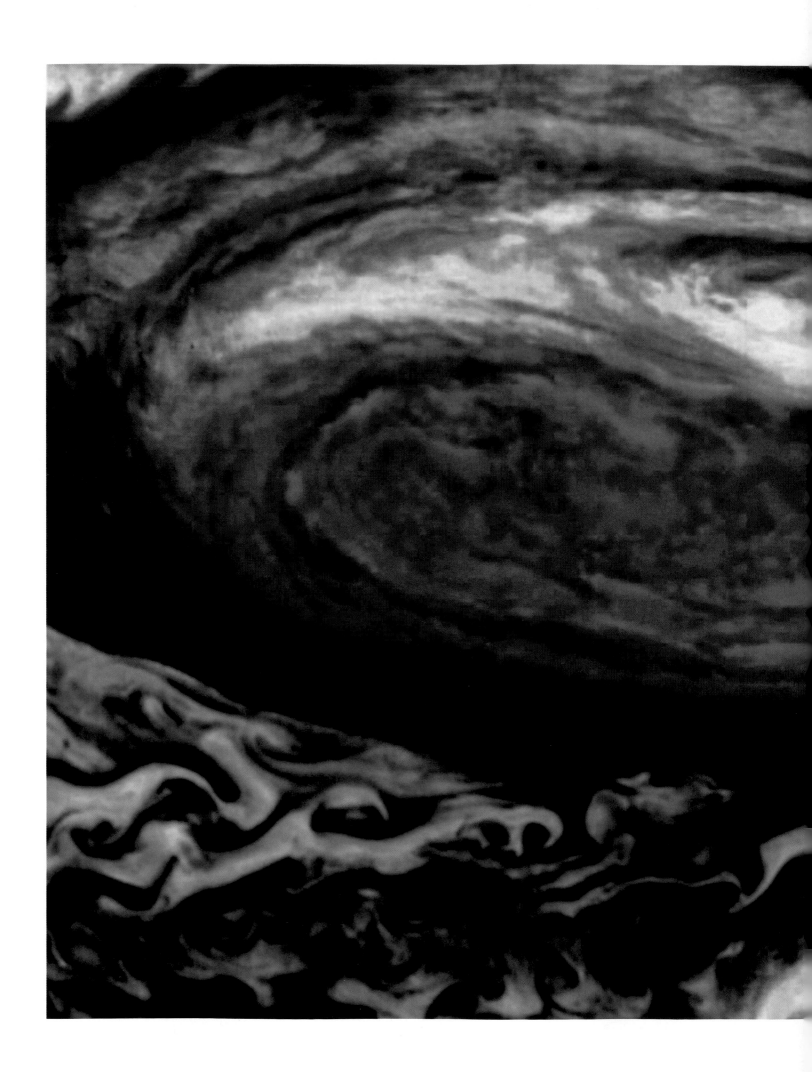

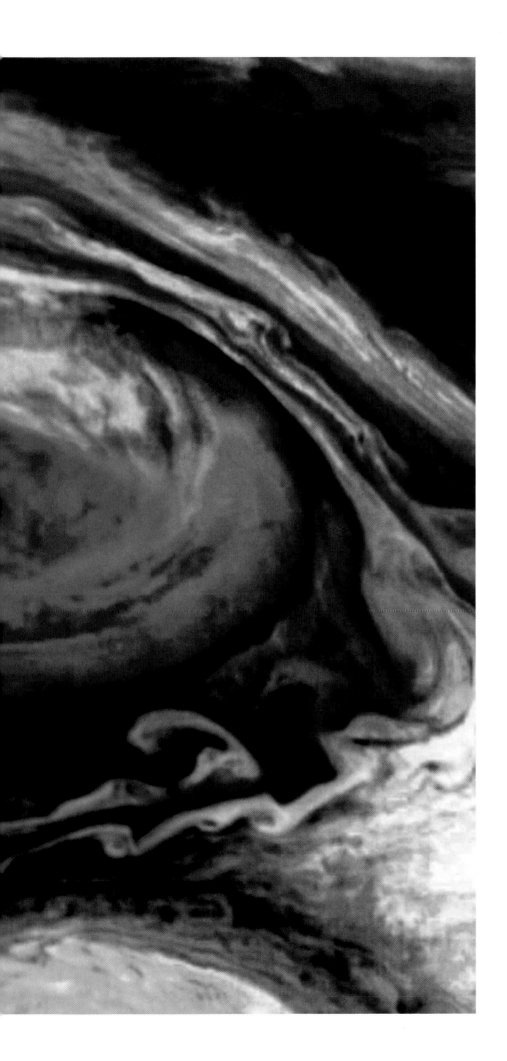

THE GREAT RED SPOT

Best-known feature of the planet, Jupiter's Great Red Spot seems to be a gigantic high-pressure storm system—two or three times wider than Earth—rotating within the clouds. It has been raging at least since the 17th century. This color-enhanced image emphasizes red and blue, and reveals a large white "wake" of vortices. The entire spot turns counterclockwise, completing a circuit about once every six days.

RINGS OF JUPITER

[FOLLOWING PAGES] Glimpsed by the Galileo spacecraft as Jupiter eclipsed the sun, the fifth planet's incredibly delicate rings consist of particles of dust, most likely spawned when meteors and high-energy particles impacted its small nearby moons. Jupiter's rings were first discovered in 1979, by Voyager 1.

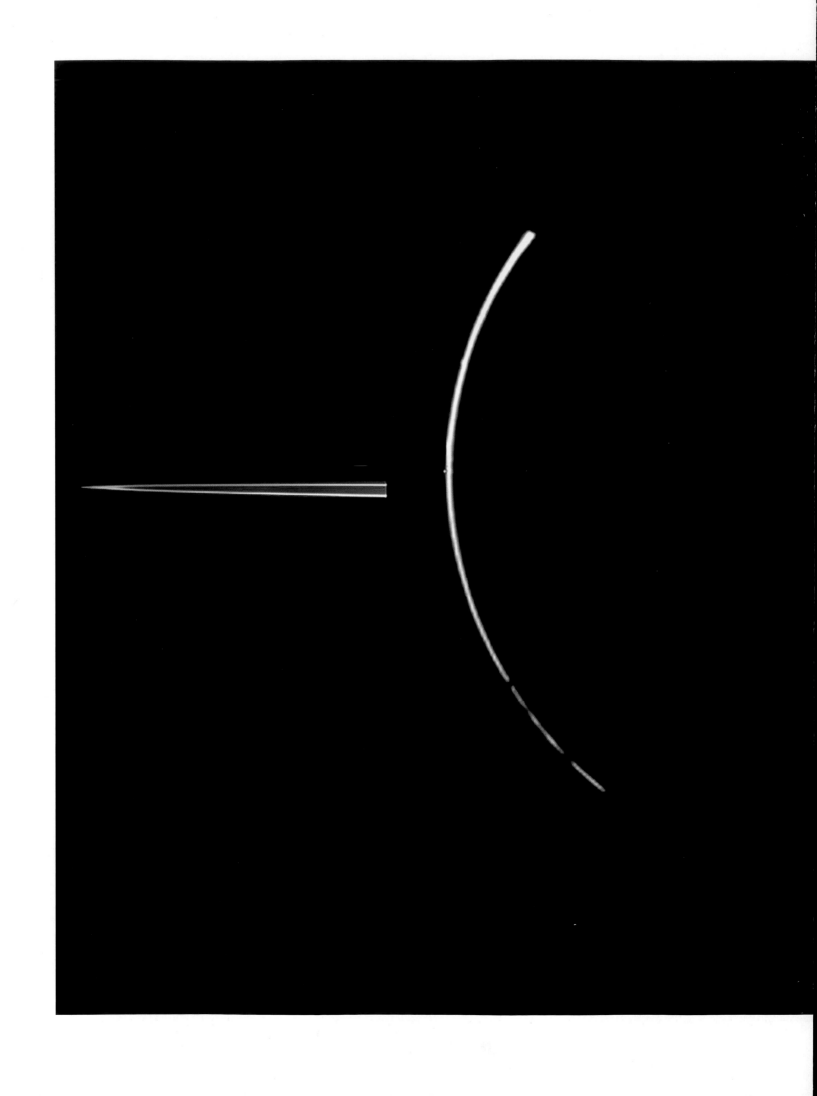

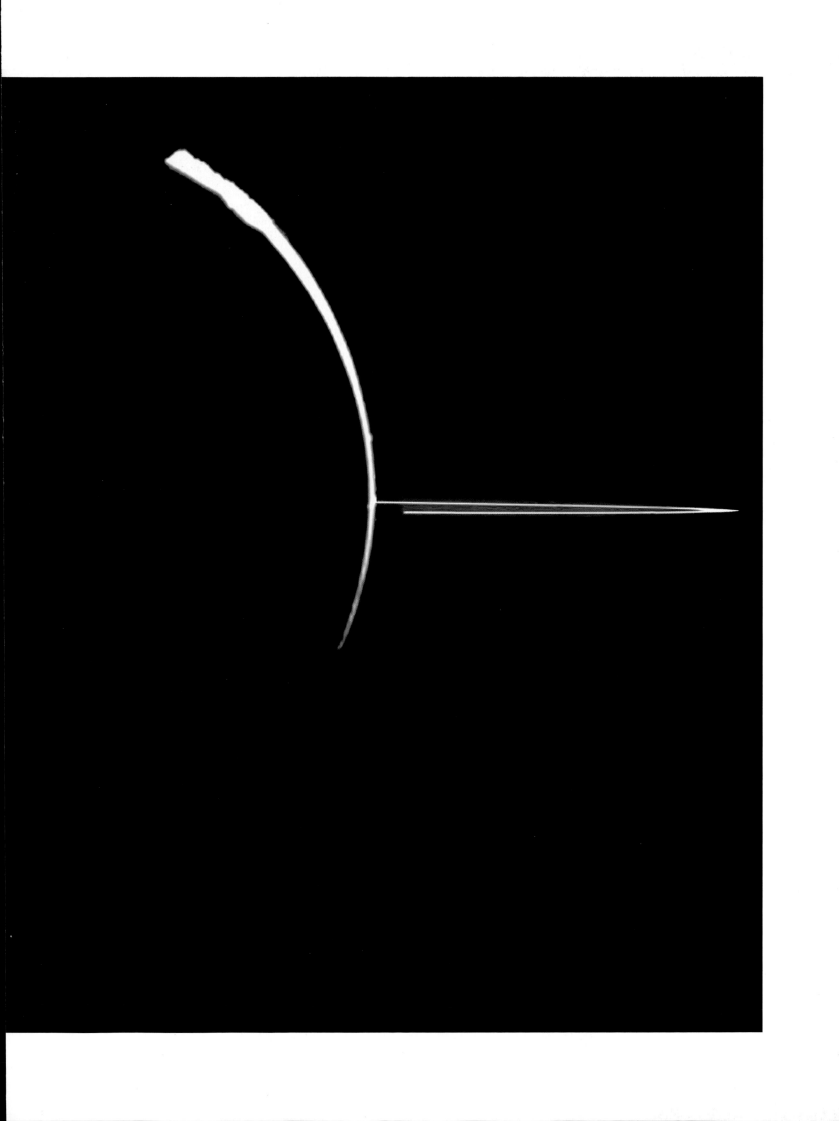

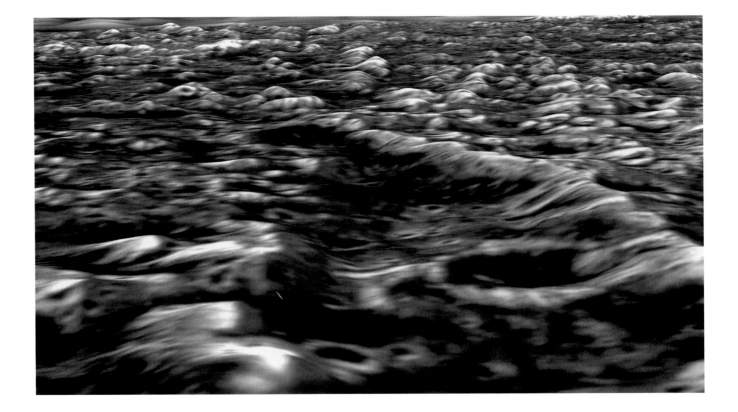

MATTERS OF SCALE

Furrowed hills of the Jovian moon Ganymede (above) rise and fall in this three-dimensional, computer-generated image, which resolves objects as small as 250 feet across. Largest moon in the solar system—bigger even than the planets Mercury and Pluto—Ganymede exhibits craters old and more recent, as well as lighter-hued regions laced with grooves and mysterious ridges.

Jupiter's innermost moon, Io, and its shadow punctuate the swirling, cloud-draped surface of their planet (opposite), visibly textured with white, reflective patches of ammonia "frost." Despite Jupiter's great size, Io takes only 43 hours to orbit it, maintaining an average distance of some 310,000 miles.

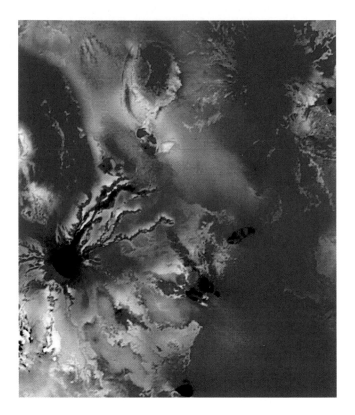

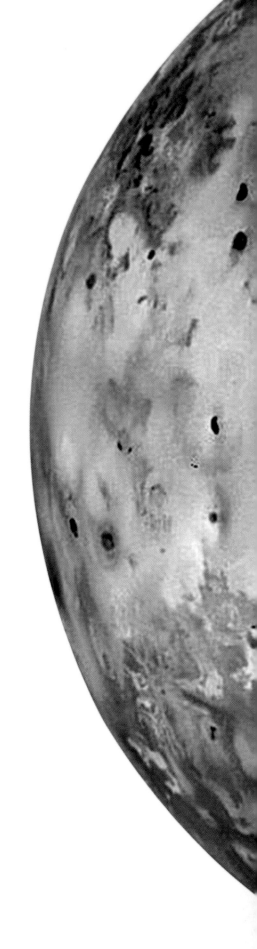

NATURAL DYNAMO

Innermost of the four Galilean
moons, Io (right) boasts the hottest
and most volcanically active surface
in the solar system, except for the
sun. Slightly larger than our moon,
it is contorted by Jupiter's nonstop
gravitational tugging, which turns its
interior into a molten inferno that
generates almost 100 trillion watts
of power in the form of heat.

Close-up of Io (above) shows red
and orange patches, signaling the
presence of sulfur compounds and
other volcanic debris that contributes
to its pizza-like appearance.

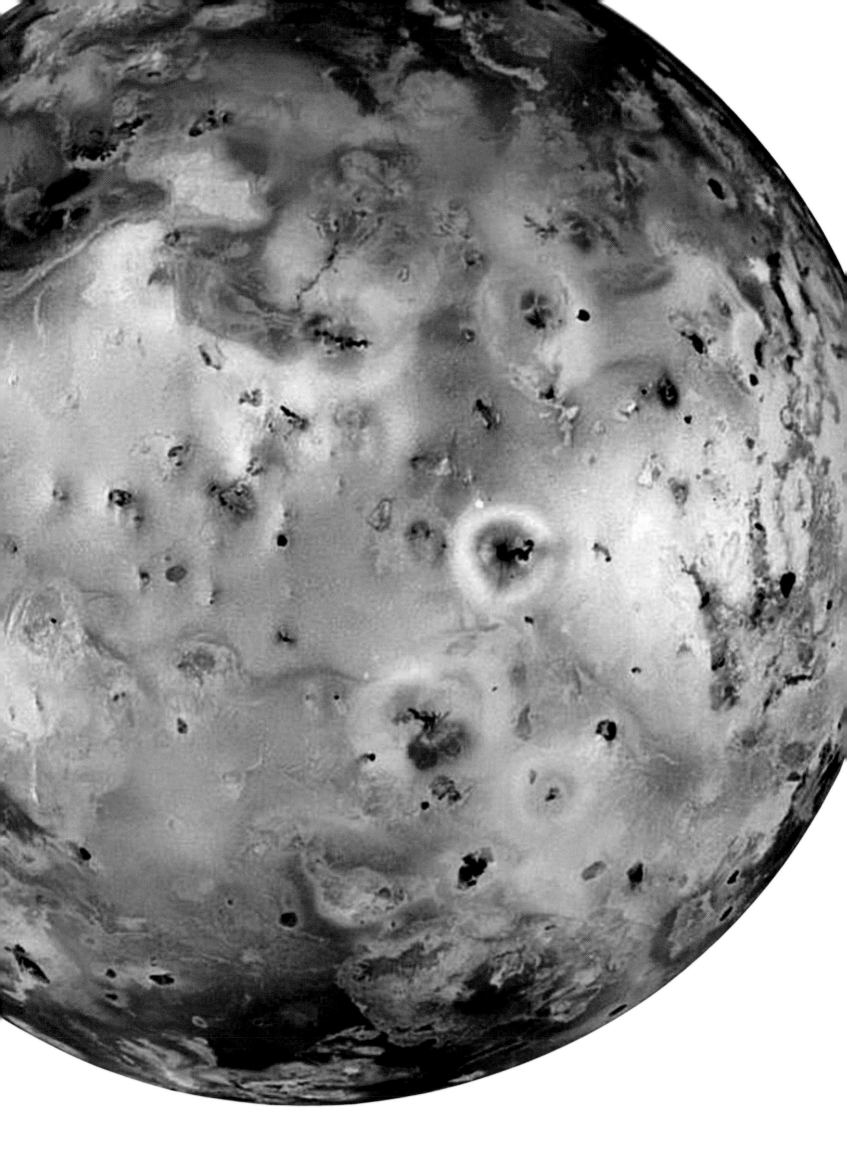

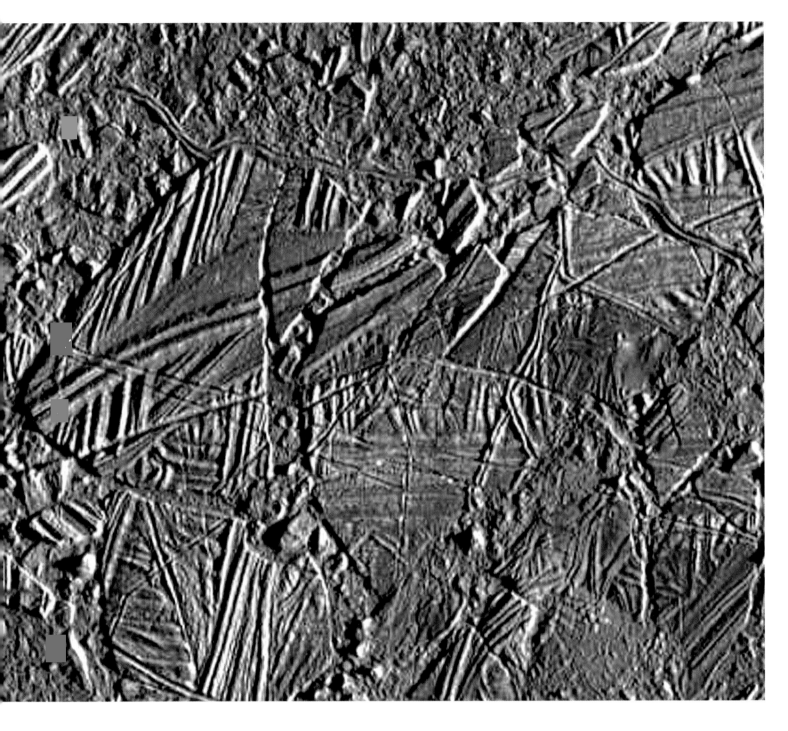

SMALL WONDER

Europa's icy crust holds the tantalizing possibility that oceans may exist beneath its sparsely cratered surface (opposite), a blend of bluish ice-plains tinged with brownish debris. Surface close-ups (above) of this smallest of the four Galilean moons reveal an eerie mix of ridges and crustal plates that seem to have a history of breaking apart and rafting together, possibly indicating subsurface water or slush.

Darker blues indicate relatively old ice; reddish regions may contain material from more recent geological activity. If large reservoirs of water do exist here, might not living organisms as well?

Currently seen as one of the best places to look for extraterrestrial life, Europa has sparked plans for a 2003 mission launch that could produce a detailed map of this moon's surface.

SATURN

Visibly flattened at the poles due to its speedy rotation, Saturn completes its day/night cycle in less than 11 hours. This infrared image from the Hubble Space Telescope reveals two fierce storms near the equator, where jet streams race by in excess of 1,000 miles per hour.

Blue indicates relatively clear atmospheric zones, as well as the presence of ammonia ice; green and yellow signal hazes of various depths; reds and oranges occur where clouds rise high into the Saturnian atmosphere.

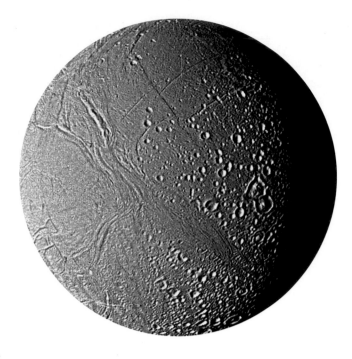

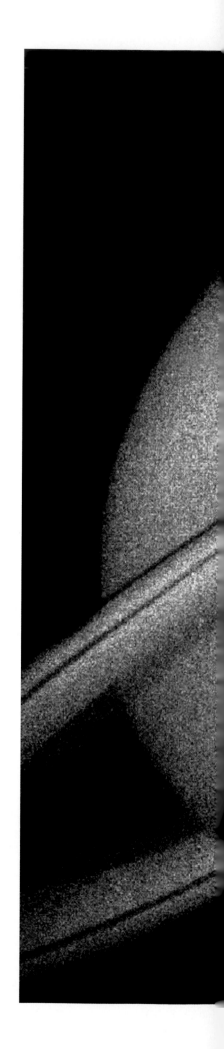

FIERY RINGS

Raging auroras crown Saturn's polar regions (right), spiking as high as 1,200 miles above the outermost clouds. Like Earthly auroras, they result when charged particles from the solar wind collide with the planet's atmosphere. Such displays glow in the extreme ultraviolet range; since our own atmospheric blanket absorbs such wavelengths, we can witness these auroras only through space-based telescopes.

Enceladus (above), one of Saturn's numerous moons, consists mostly of frozen water and is considered to have the purest ice surface in the solar system. Grooves and a relative lack of craters indicate its crust is relatively young, causing some astronomers to believe it may be volcanically active.

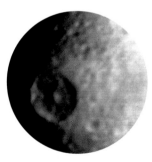

A dirty snowball by comparison, Mimas (left) contains ice and traces of rock. Its huge signature crater, named Herschel, honors the discoverer both of this moon and Enceladus, English astronomer Sir William Herschel.

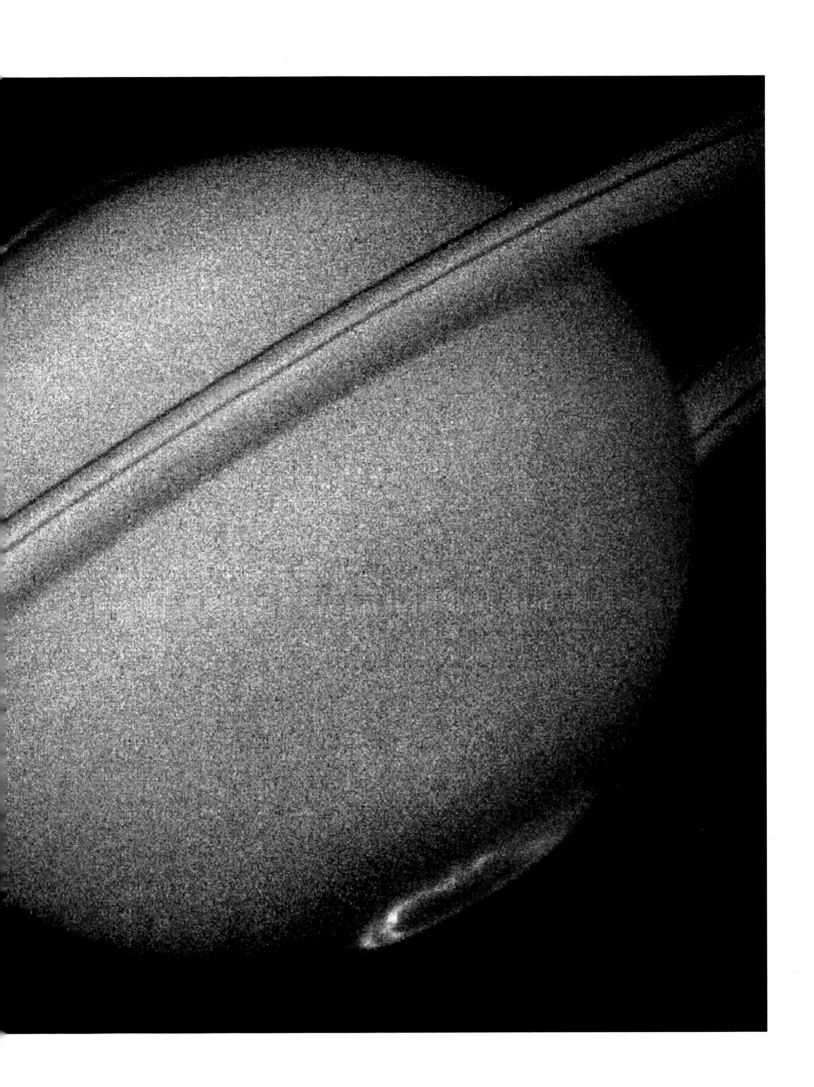

SATURN'S GLORY

Color-coded to enhance subtle differences, a Voyager 2 image of Saturn's brilliant rings (opposite) shows that—although originally thought to be a single broad band, then three lesser ones—the rings actually make up thousands of concentric "ringlets."

Each is composed of ice particles that can be as small as sand grains or as large as houses, speeding along at up to 500,000 miles per hour. The elaborate ringlets-and-gaps structure has been explained largely by the complex gravitational effects of this planet's numerous moons and even smaller "moonlets" embedded within the rings.

Seen edge-on (above), Saturn's rings shrink to spidery thinness while Titan, Saturn's biggest moon, floats just above them at left, casting its shadow on the cloud-draped planet. On the right, four other moons cluster: Mimas, Tethys, Janus, and Enceladus (left to right).

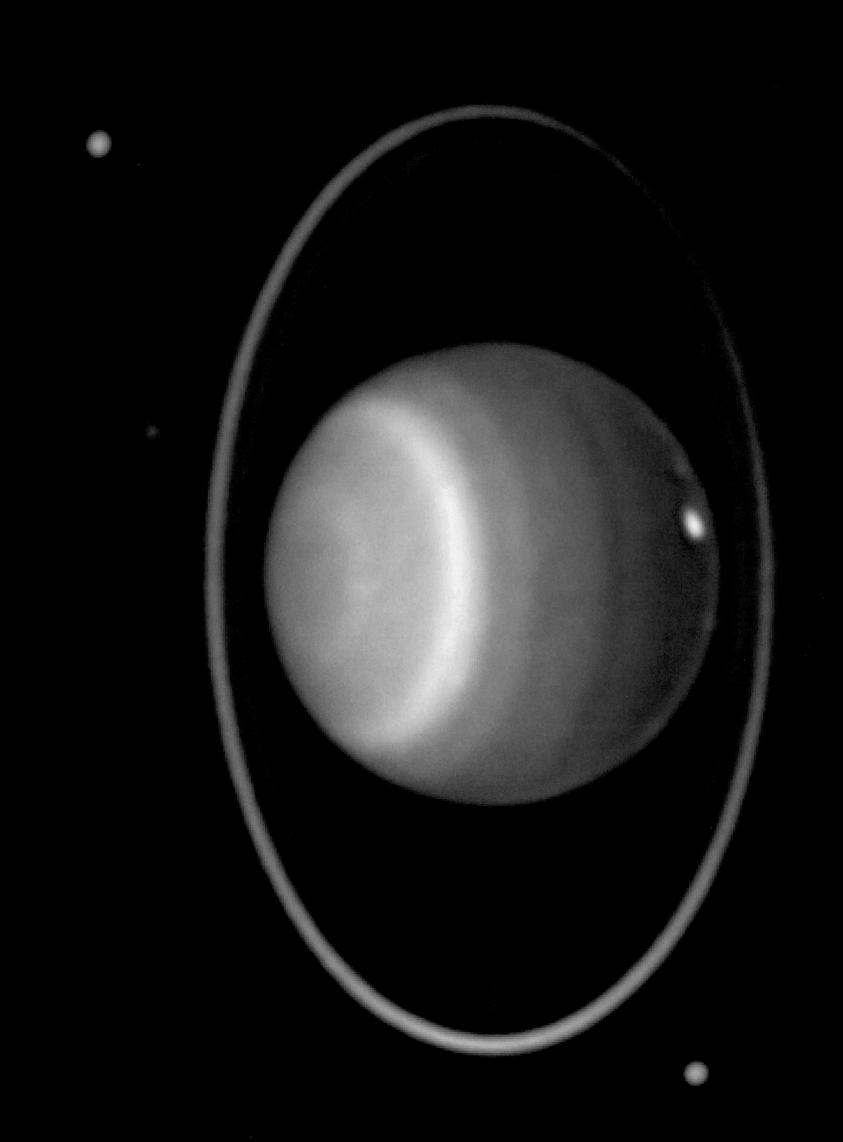

URANUS

Recent, false-color view of the seventh planet as seen by the Hubble
Space Telescope reveals four major rings, ten of the seventeen known
moons, and atmospheric information. Green and blue regions identify
clear areas where sunlight can penetrate; yellows and grays indicate
a haze or cloud layer causing some reflection of light. Oranges and reds
signify very high cloud cover, similar to cirrus clouds on Earth.

 The rings of Uranus are thin, narrow, and dark compared to those of
other gas giants. Their component ice particles have been so darkened by
rock debris that they reflect light about as poorly as charcoal. Yet they are
speedy, each taking only eight hours to complete a round-trip of the planet

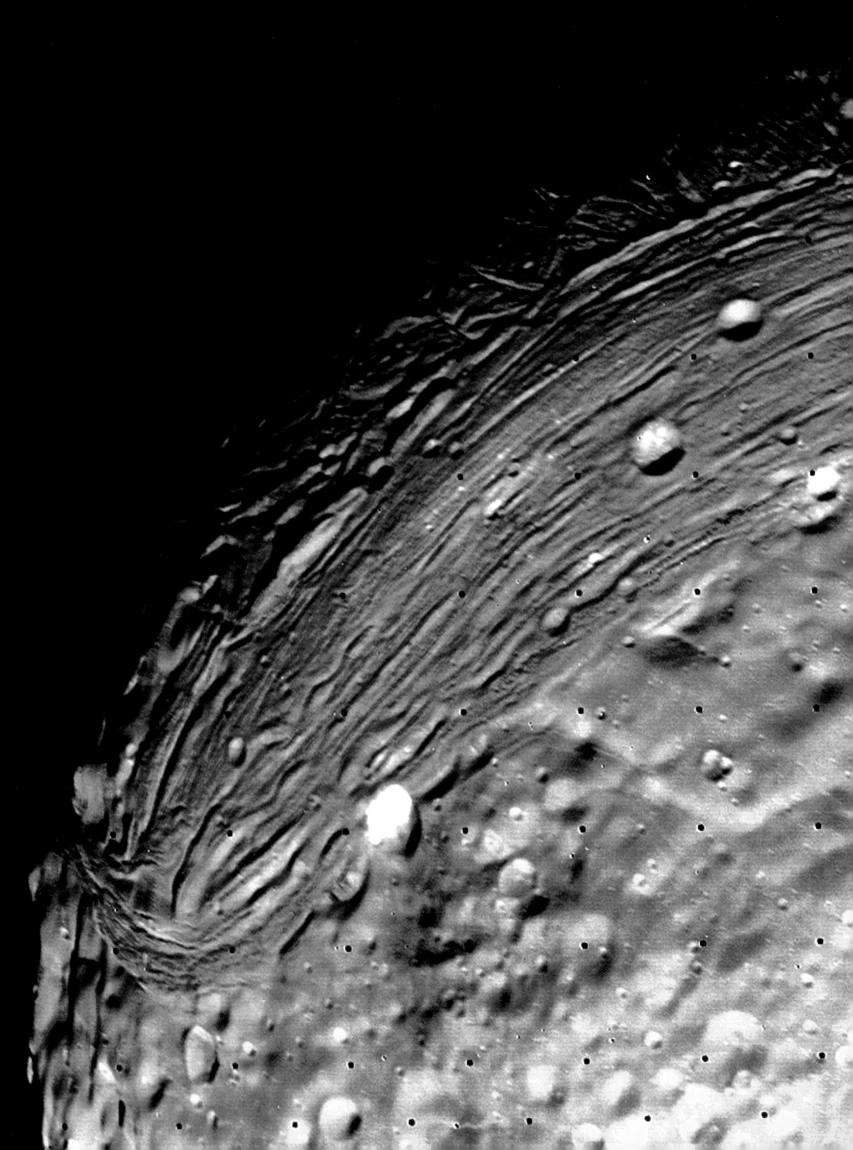

VIEWS FROM VOYAGER 2

At a distance of 1.7 million miles, the only spacecraft to have probed the Uranian system produced this false-color image of the planet in 1986 (below). High-altitude hazes show up as pink; blue regions are the most cloud-free. Actually Uranus is blue-green in color, due to absorption of red light by methane gas in its deep, cold, and remarkably clear atmosphere.

The International Astronomical Union has dubbed this planet's moons and their surface features with Shakespearean names. Thus, a close-up of the moon Miranda (opposite) focuses on a grooved region known as Elsinore Corona, evidence for vigorous tectonic activity in the past. Similarly, one of the planet's more distant moons, Oberon (above), honors the fairy king from *A Midsummer Night's Dream*; this Oberon, alas, is tragically flawed by a crater named Hamlet.

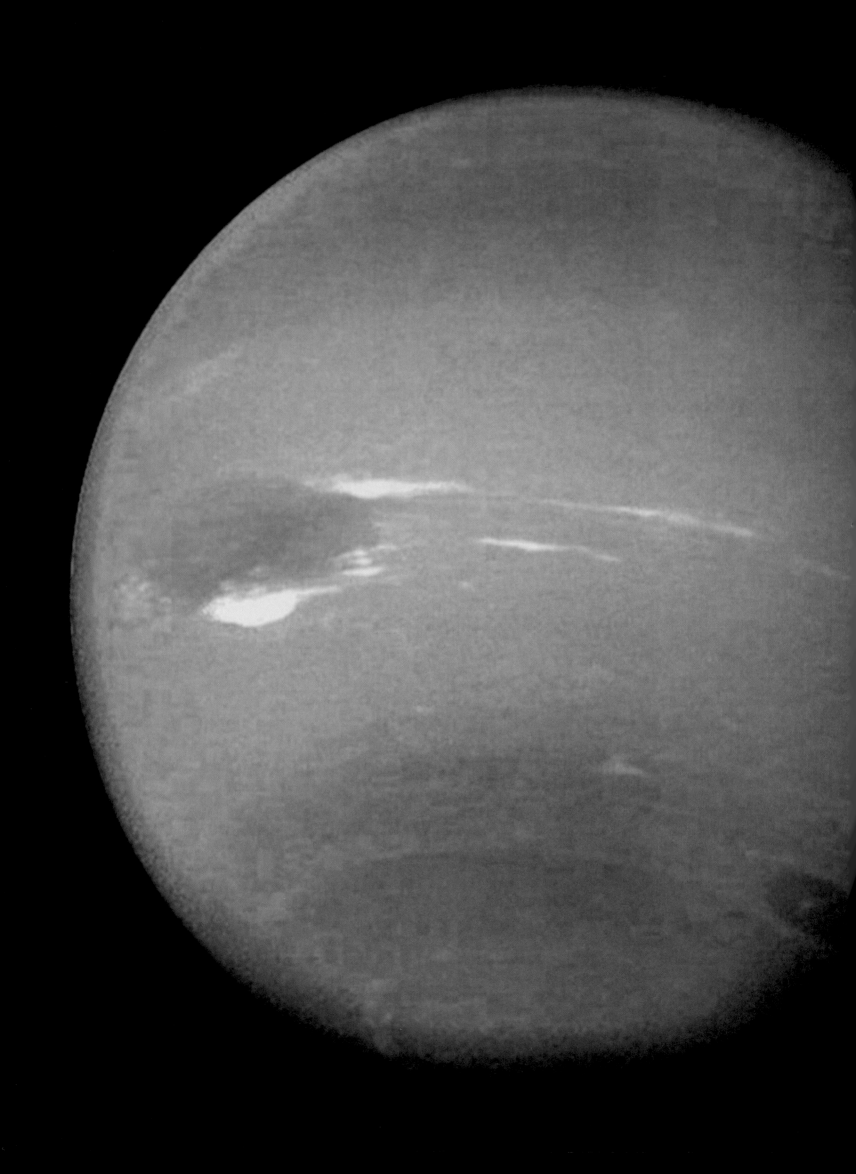

NEPTUNE

Thirty times farther from the sun than Earth is, distant Neptune receives only 3 percent of the sunlight that Jupiter does—and a minuscule fraction of what reaches us. Yet it is a dynamic world. Its Great Dark Spot, a massive rotating system seen here, compares in size and southern latitude to Jupiter's Great Red Spot but is much less stable. When the Hubble Telescope viewed it in 1994, the spot had vanished—only to be replaced by another one in Neptune's northern hemisphere!

PARTING SHOT

After flying by Jupiter, Saturn, Uranus, and finally Neptune, Voyager 2 turned back and bade farewell to the outermost gas giant (left). The spacecraft also sent back data of Triton, Neptune's largest moon by far, about 1,700 miles in diameter.

Beneath a thin atmosphere, Triton's dark, tortured surface (opposite) harbors some unusual morphological features, as well as evidence for ice volcanoes. Prominent dark streaks, visible here, seem to originate from small volcanoes and may consist of nitrogen frost mixed with organic compounds ejected during geyser-like eruptions. Surface temperatures, however, average nearly minus 210°C.

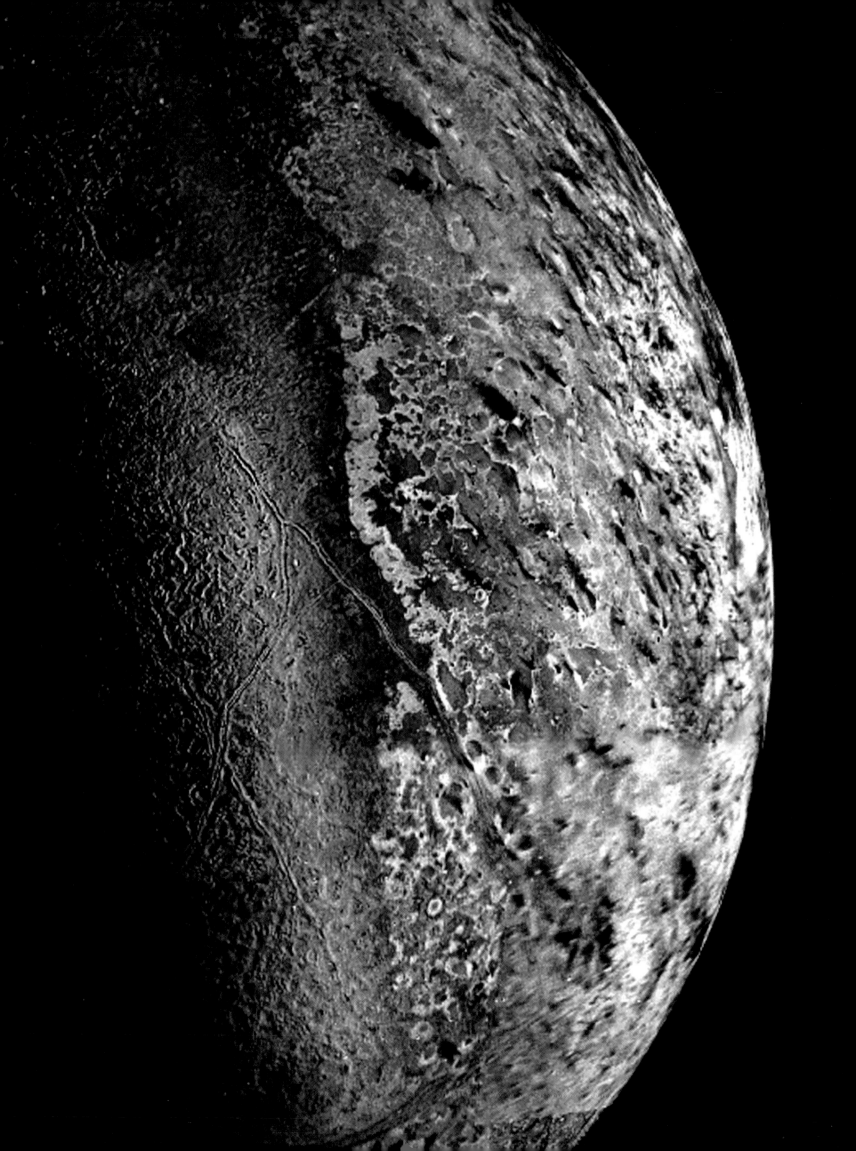

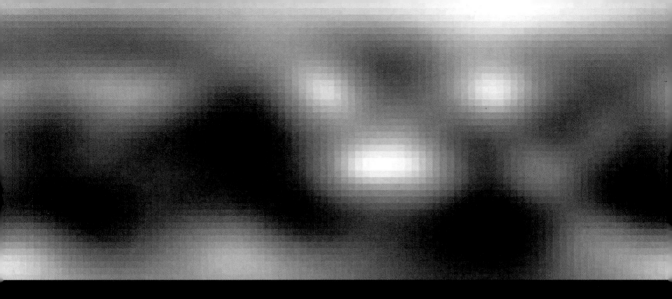

Dim and fuzzy due to their extreme distance from Earth, Pluto and its single
moon, Charon, (right), qualify as the least visited planetary group in the solar
system. No landers or flybys have ventured to either body—but the Hubble
Space Telescope has enabled astronomers to project a shadowy, Mercator-
like map of the planet's entire surface (above). Images chronicle Pluto's
6.4-day rotation cycle and reveal its dark equatorial belt, bright polar caps,
and certain physical variations that may be seasonal or topographic.

Astronomers have calculated Pluto's diameter at about 1,430 miles—
roughly two-thirds the size of our moon (and 1,200 times more distant).
Charon, roughly 780 miles across, has somewhat bluer coloration, suggesting
that it differs in surface composition and structure from its home planet.

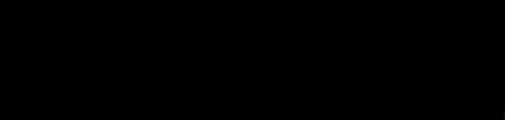

COMETS

Perhaps the most viewed comet in history, Comet Hale-Bopp
takes center stage in the constellation Cygnus during 1997.
So bright that it was visible to the naked eye even in light-
polluted cities, this comet easily competed with near-twilight
skies and even the crescent moon.

Where is it now? Still visible to telescopes in our Southern
Hemisphere, the outbound comet is presently about 537 million
miles from the sun. (Jupiter orbits at about 480 million miles.)
According to calculations, Hale-Bopp should return to Earth in
about 4,200 years.

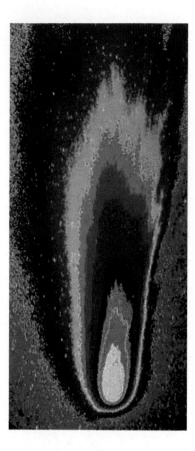

SURREAL "SNOWBALLS"

False-color images of Comets
Halley (above) and Hale-Bopp (right)
indicate relative brightness of their
various parts. Hale-Bopp, here
some 340 million miles from Earth,
eventually came as close as about
120 million miles. Stars—far more
distant—appear through the comet's
expanding coma, a cloud of gas and
dust generated as its icy nucleus
sublimates due to solar heating.

Each time Halley returns, its
nucleus sheds about a 20-foot
thickness of ice and rock into space,
creating its noted tail. Billions of
comets are thought to orbit our sun,
but most remain too far from Earth
to be seen.

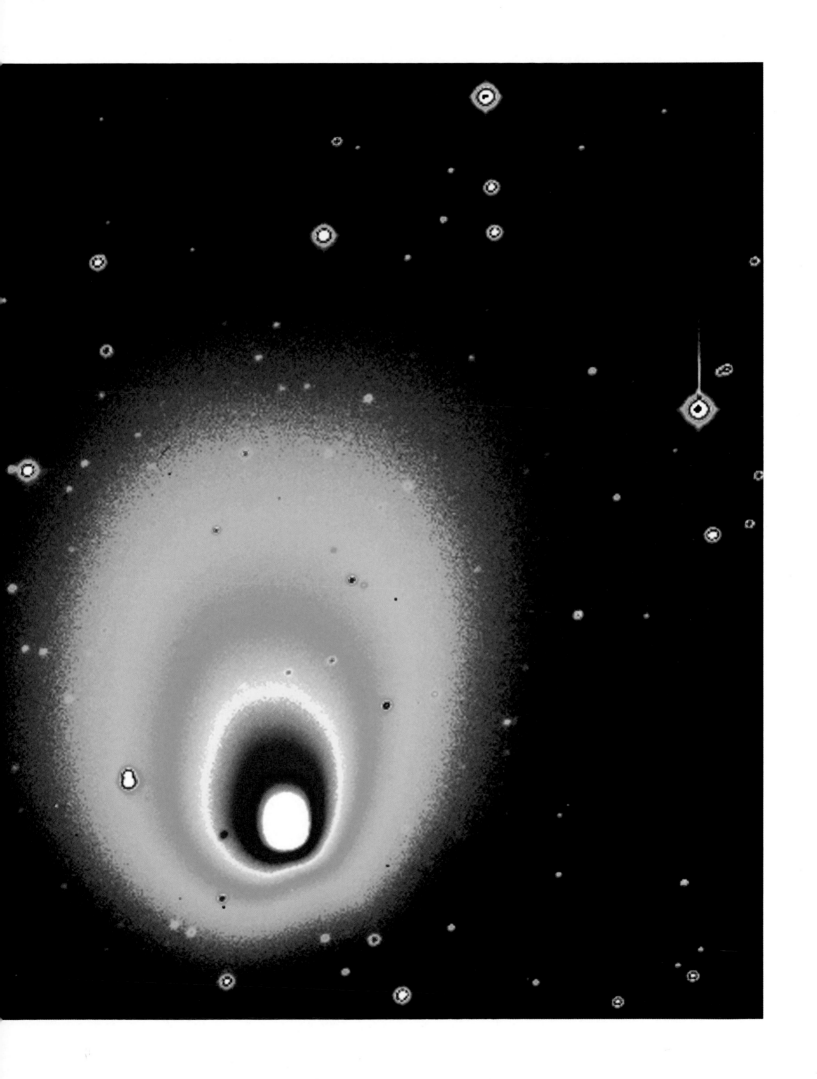

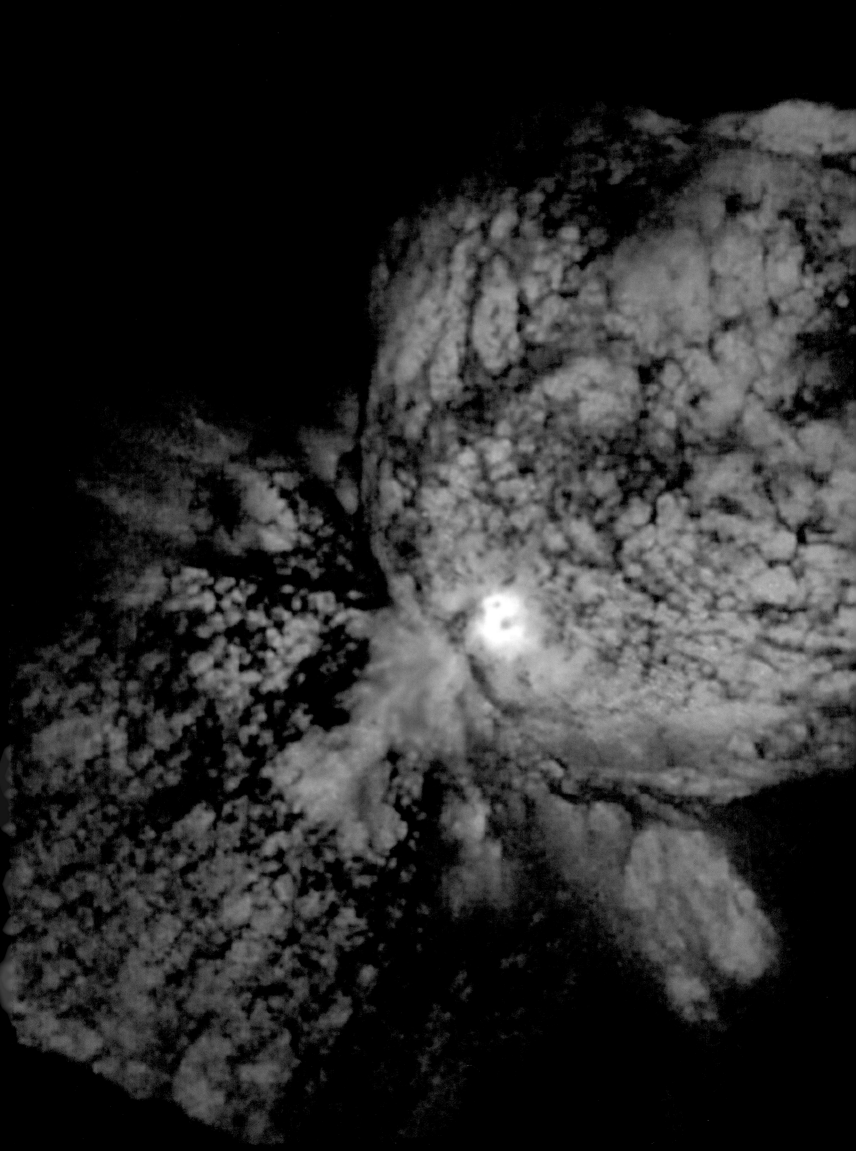

THE GREAT
BEYOND

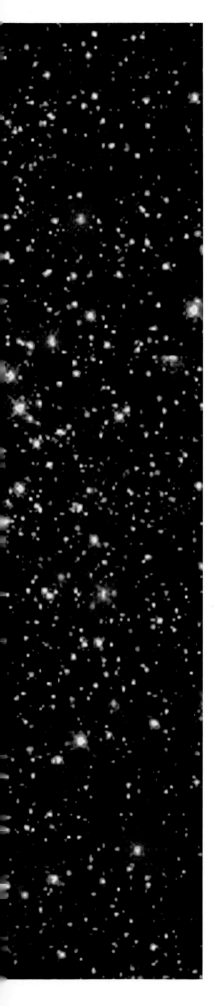

GIVEN THE SPLENDOR AND DIVERSITY OF OUR SOLAR SYSTEM, it is sometimes hard to believe that all these wonderful planets and moons orbit one rather ordinary star living its life in a very ordinary section of a very ordinary galaxy. Yet that is what 20th-century astronomy has taught us about our place in the universe. The minute that we look beyond our own solar system, we find another, almost unbelievable realm of stars that could swallow our sun without a ripple, of black holes sucking in every speck of matter around them, removing it forever from the known universe. It is a realm where giant stars end their lives in titanic explosions, throwing chemical elements into space—elements from which planets like Earth can eventually be formed. It is a realm of much greater size and diversity than anything that we have experienced on our tour so far. It is to this realm—this "great beyond"—that we now turn our attention. Just remember: It's a realm where truth often is stranger than fiction.

It took astronomers a long time to find out just how great distances can be when we leave our own home system. The reason is that the night sky is two-dimensional—the tiny point of light that reaches us from a star or planet gives no indication as to how far it has traveled. Is it a dim star that happens to be relatively near, or is it a very bright star that's far away? Without adding a third dimension to the flat night sky, there's no way to tell. That's why, although the stars in a given constellation appear to be connected, the distances of the individual member stars from Earth actually can—and often do—vary immensely. Providing depth to our two-dimensional sky has turned out to be one of the most difficult and enduring problems in astronomy. In fact, that task is not yet completed.

The basic tool at the astronomer's disposal is triangulation—measuring the angles to an object from two different places and then, by knowing the distance between those places, deducing the distance to the object. (The distance between the two observation points is called the baseline.) Greek astronomers in the second century B.C.,

FESTIVAL OF STARS

Two relatively young star clusters blaze from within our nearest galactic neighbor, the Large Magellanic Cloud. Yellowish lights signify 50-million-year-old main-sequence stars, which—like our sun—subsist by fusing hydrogen into helium. White stars are much hotter, more massive, and only about four million years old.

[PREVIOUS PAGES]
Doomed star Eta Carinae, roughly a hundred times larger than our sun, burns millions of times more brightly, billowing twin lobes of dust and gas from its hot central region. Eventually it will explode in a mighty cataclysm we call a supernova.

for example, used this basic technique to determine the distance to the moon. But triangulation has a fundamental limitation: It depends on the astronomer's ability to measure angles accurately. If you want to determine the distance to a far-off object by measuring from two different places on Earth, you must be able to establish that the two angles of sighting are, in fact, different from each other. If you can't, then as far as you can tell the two lines of sight are parallel—and the object is an infinite distance away. There are only two ways to get around this difficulty. One is to make the difference between the angles greater by increasing the distance between the observation points; the other is to find a way to measure angles more accurately.

Before the invention of the telescope in the early 17th century, astronomers had only the first option available. There is, of course, a limit to how far apart two observation points on our planet can be—they can't be more than one Earth diameter apart. (In practice, you wouldn't actually rely on two observers at opposite sides of the planet; it's much easier to take two measurements with the same instrument at the same place on Earth twelve hours apart. This way, Earth's daily rotation provides the two different observation points.) By measuring the angles of sight to the moon at different times, Ptolemy made a good estimate of the Earth-to-moon distance by A.D. 150. Unfortunately, it turns out that with naked-eye instruments and a baseline of one Earth diameter, it is not possible to measure distances to anything else. Determining the distance from Earth to the sun or planets had to wait until the telescope not only had been invented, but also had been considerably refined.

In 1672, scientists made a breakthrough. Using state-of-the-art telescopes—which increased the accuracy of the measurement of angles—and a newfound precision in measuring locations on Earth's surface (thus providing more accurate baselines), they succeeded in determining the distance from the Earth to the sun—what we now call the astronomical unit, or AU. Now the stage was set for a massive increase in the available baseline; instead of being limited to the diameter of the Earth, scientists could now use the diameter of Earth's *orbit around the sun* as their baseline. Theoretically, at least, they could measure the angle of the line of sight to a star in July and January, for example, then triangulate to determine that star's distance. Unfortunately, stars are so very far away that the increased baseline alone wasn't enough—astronomers had to wait over a century before telescopes improved to the point that they could detect differences in the angular measurements to a given star.

In 1838, German astronomer Friedrich Bessel became the first to determine a star's distance from Earth by triangulation. He chose 61 Cygni, in the constellation Cygnus (the Swan), and after about 18 months of hard work announced that it was 690,000 AU, or roughly

NEWBORN NEBULA

Like a cosmic bird in flight, the extremely young binary star named CoKu Tau/1 puts forth nebulous "wings" of light that stretch as far as 75 billion miles into space, each outlined by light-reflecting dust particles. At its core, two stars anchor this nebula, about 450 light-years away, in the constellation Taurus.

10.9 light-years from Earth. This is a huge distance. A spaceship that could speed across the 93 million miles from the Earth to the sun in a single hour, for example, would take over 78 *years* to get to 61 Cygni. With Bessel's measurement, the universe suddenly became far larger than humans had ever imagined.

In time, the distances to other stars were determined as well. But as the apparent size of the universe grew, the sheer magnitude of the distances involved defeated even the best telescopes and the most careful attempts at triangulation. To map the great beyond, scientists had to find another distance-measuring tool. It wasn't until the early 20th century that American astronomer Henrietta Leavitt, working at the Harvard Observatory, laid the groundwork for a new technique, which goes by the name of standard candle.

IN THE JARGON OF COSMOLOGISTS, A STANDARD CANDLE IS an object in the sky whose energy output is known. A 100-watt lightbulb would be a perfect standard candle, since we know that it generates power at the rate of 100 watts. Its usefulness is this: If you know how much energy it's giving off and how much energy you're actually receiving on Earth, you can figure out how far away it is.

For decades, astronomers hoped to find some object in the sky whose brightness could be determined by an observation that didn't require knowing its distance—in effect, finding a way to accurately read the label on the lightbulb, then using the energy output of this standard candle to measure its distance. But where were the wattage signs in space? Leavitt's work centered on a type of star known as a Cepheid variable (so named because the first such star studied was in the constellation Cepheus, the King). It is a star type whose output isn't constant, but changes with time.

If you watch a Cepheid variable over a period of many months, you notice that it brightens and dims repeatedly, according to a regular cycle. Leavitt was able to establish that the total amount of light each Cepheid variable radiated into space was related to its period of brightening and dimming—the longer the period, the brighter the star. This was a major breakthrough in astronomy, because it meant that if astronomers could find a Cepheid variable in a star cluster, they could watch it for a while to determine its period, and from this information they could deduce the total amount of energy it was radiating into space. By comparing this to the amount of light actually received on Earth, they could then calculate the distance to that variable, and hence to all the stars in its cluster.

It was by using Leavitt's standard candle that astronomers early in this century began to map the great beyond. They found that the stars of the Milky Way were arranged in spiral arms, and that the sun was not at the center of the universe, but off to one side. They also

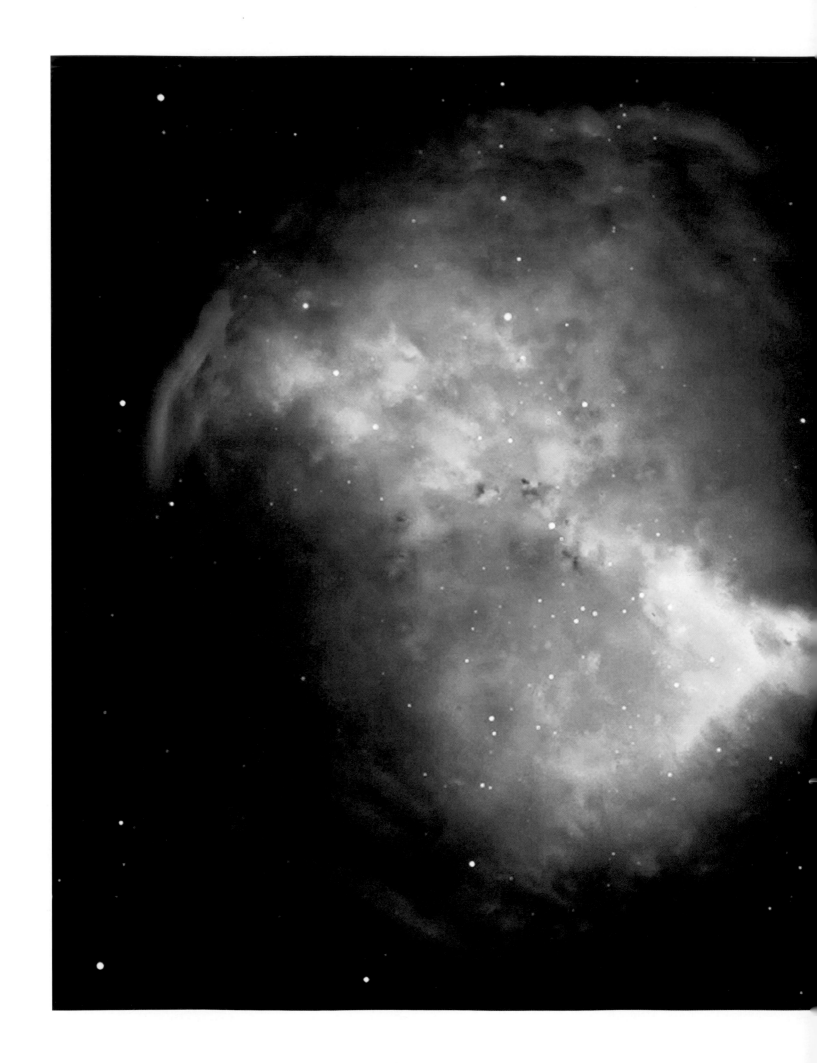

determined that the farthest individual stars they could see traced out a Milky Way about 300,000 light-years across. (We now know that 100,000 light-years is more accurate.) Throughout all this exploration there remained the puzzle of the nebulae.

"Nebula" is Latin for "cloud," and astronomers first used the term as a generic catchall for fuzzy, cloudlike objects that could not be resolved by even the best telescopes until well into the 20th century. Debates raged for decades as to whether these patches were simply clouds of gas and dust within our own Milky Way, or were actually—in the language of the times—other island universes, vast galaxies of stars like the Milky Way but much more distant.

The problem was that, when viewed through telescopes of low resolving power, many types of distant objects tend to look the same. They're simply fuzzy patches of light in the visual field. It was only when telescopes improved to the point that they could resolve the structural details of nebulae that the debate itself was resolved. As we shall see, nebulae turn out to be some of the most beautiful and awe-inspiring objects in the heavens.

In 1925, a rather extraordinary individual entered the nebulae debate. Edwin Hubble was born in Missouri and attended the University of Chicago, where he played championship basketball and became so skilled a boxer that he supposedly considered a career in the ring. After serving in World War I, he returned to Chicago for a law degree, and tried lawyering for a year. He then went back to Chicago, this time for a degree in astronomy. Soon after, he became one of the first astronomers to gain access to the new 100-inch telescope on Mount Wilson, near Los Angeles.

With the light-gathering and resolving power of this telescope, Hubble was able to make out individual stars in many nebulae, and then apply Leavitt's data to determine their distance from Earth. In particular, he showed that many nebulae are not just thousands but millions of light-years away. Since our own Milky Way was then thought to be 300,000 light-years across, that meant that these nebulae in fact constituted their own galaxies, or island universes, giving us a context within which to study the Milky Way.

Not all nebulae, however, are galaxies. With the help of modern telescopes, we have been able to discern several different nebular categories. There are, for example:

PLANETARY NEBULAE. When a star like our sun reaches the end of its life, it goes through a brief period as a red giant, expelling enormous amounts of material into space in a kind of greatly magnified solar wind. Some time later, astronomers looking at the core that's left will see a glowing center surrounded by a much larger shell of gas. Viewed through low-resolution telescopes of

PLANETARY NEBULAE

Though they often appear round and planetlike even when seen through a small telescope, so-called planetary nebulae have nothing to do with planets. They chronicle a stage in the life of stars similar to the sun; after such a star becomes a red giant, it expels its gaseous envelope, providing an often dramatic display that fades over thousands of years—while the star becomes a white dwarf. Some expelled envelopes display exquisite symmetry, as occurs in planetary nebula M27 (left), also known as the Dumbbell.

the 19th century, such shells often appeared to be the sort of disks in which planets might be forming, much like the protoplanetary disk that we believe gave rise to the planets in our solar system. For historical reasons, astronomers still refer to these as planetary nebulae, even though we now know they are stellar events, and have nothing to do with planets. They remain some of the most visually intriguing objects, often exhibiting shapes that vary from dumbells to rings and spirals to angular, almost polygonal forms.

SUPERNOVA REMNANTS. When stars that are much larger than our sun die, they do so spectacularly, in titanic explosions called supernovae. The giant star's outer layers are ripped off and hurled into space. In time, this distant but expanding cloud of material appears to our eyes as a fuzzy patch of light. Perhaps the best known nebula of this type is the Milky Way's Crab Nebula (pages 214-215), the glowing remains of a supernova that exploded nearly a thousand years ago.

EMISSION NEBULAE. Most nebulae inside the Milky Way are actually huge clouds of interstellar dust or gas, unassociated with any particular star. The most dramatic of these clouds are called emission nebulae, and they are so hot that they give off light. A good example is the Great Nebula of Orion (pages 202-203), where hot young stars heat clouds of surrounding hydrogen gas to some 10,000°C—about twice the temperature of the surface of our sun—by absorbing ultraviolet light from those stars. The clouds then emit light themselves, which shows up in our telescopes. Hydrogen is particularly good at emitting such light. Sometimes, dust mixed in with the hot gas absorbs some light, creating darker regions that contrast dramatically with the rest of the nebula.

DARK CLOUD NEBULAE. As we've seen with emission nebulae, clouds of interstellar dust often absorb light, creating dark areas. The black band that seems to split the Milky Way, for example, is not caused by a lack of stars in a particular region, but by the fact that light from the stars that are there is absorbed by intervening clouds of dust. The Horsehead Nebula (left, and pages 204-205) is one well-known example.

REFLECTION NEBULAE. Finally, there are nebulae that do not shine with their own light, but are dark. Even so, their dust reflects light from other stars that do not heat the nebulae enough to make them glow. Reflection nebulae often appear to be bluish in color, because dust scatters light at the blue end of the spectrum more readily than it does at the red end. In fact, this also explains why, whenever the sun shines, the Earth's sky appears blue to its inhabitants.

DARK CLOUD NEBULAE

One of the best-known images in astronomy, the Horsehead Nebula's namesake profile is part of a much larger dust cloud. Like other dark cloud nebulae, it seems dark due to absorption of light by the dust— while starlight happens to illumine this nebula's background, causing hydrogen there to fluoresce and thus backlight the "head."

THE MILKY WAY GALAXY CONTAINS A FEW HUNDRED billion stars, only a few thousand of them visible to the naked eye. Yet it would be wrong to think of this galaxy as just stars; it is much more complicated. Nevertheless, stars are certainly one of its most striking components. Like our sun, they arose from collapsing clouds of interstellar gas and dust, and managed to counter the pull of gravity and stave off their collapse by causing hydrogen to fuse into helium in their cores. Every star, from the largest to the smallest, begins life as a glowing ball containing primordial hydrogen, which is consumed to produce the energy that the star needs. All stars are not the same, however. Depending on the density, temperature, and structure of matter in the cloud from which it sprang, a star can be much smaller than our sun or much larger. In fact, if you imagine the sun as being the size of a basketball, we know of stars in the sky as small as BBs and as big as 50-story buildings.

In addition, some stars are bright and some are dim, some are faintly bluish while others are white, and so on. Apart from huge variations in their distances from the Earth, stars differ from each other in two basic ways: They have different masses (that is, different amounts of material in them) and they are at different stages in their life cycles.

Looking at stars in the night sky is similar to looking at a vast and diverse forest that not only has many different kinds of trees—oak, birch, maple, pine, and so on, but also includes trees of many different ages. Some are only seedlings, some are young, some are mature, and some are dead and rotting. In the same way, the Milky Way is diverse because its stars—like the stars of any galaxy—vary greatly in mass and in their stellar stage of life.

By the end of the 19th century, astronomers knew that many kinds of stars existed and were beginning to classify them according to the light they emitted. By 1920 they found that they could gauge a star's surface temperature by determining the brightness of the frequencies of light that are emitted only by atoms that have lost one or more electrons. The more electrons lost, the higher the star's temperature. Today, surface temperature and brightness are used to classify stars on a scale that ranges from hot to cool, using the letters O, B, A, F, G, K, and M. (Generations of astronomy students have memorized the sequence through the line, "Oh Be A Fine Girl/Guy, Kiss Me.") Our sun is a G-type star, medium both in size and temperature. Scientists believe that such stars are ideal for planets such as Earth, because they make it possible for liquid water to remain on some planetary surfaces for long periods of time, a prerequisite for life, as we know it, to have developed.

You might expect that the bigger the star is, the longer it will last—after all, a larger star has a lot more hydrogen, a lot more fuel to burn. As it turns out, however, the situation is more involved. For although

a big star has more hydrogen, it also exerts a larger gravitational force on its components; therefore it has to burn its fuel more rapidly to keep from collapsing. Because of this, we have the seemingly paradoxical result that the more fuel a star has, the shorter its lifetime. Our sun is a medium-size star that started out with enough hydrogen to burn for about 11 billion years. A star 40 times larger may live only a little over a million years, while a star with half the sun's mass can be expected to putter along for two hundred billion years as it frugally expends its supply of hydrogen. The watchword for stars seems to be: Live fast, die young—and make a spectacular corpse.

Four-and-a-half billion years ago, when the sun condensed from a cloud of interstellar dust and gas and first began to shine, that dust cloud would have absorbed much of its visible light. But as the fusion reactions in its core increased in strength and the dust cloud dissipated, the sun shone brighter and brighter, quickly becoming almost as bright as we see it today.

Stars like the sun, which are still burning hydrogen in their cores, are called main-sequence stars. One of the most interesting questions about them concerns what happens when they run out of hydrogen fuel. Our sun, for example, will run out of hydrogen in its core in about six billion years. At that point, its outpouring of energy will begin to diminish and will no longer counterbalance the force of gravity, which will have been waiting in the wings for 11 billion years. Material will be pulled in and the sun will start to contract. This contraction, in turn, raises inner temperatures, so that any hydrogen left in layers outside the core will begin to burn. More important, the helium at the sun's very center—the "ashes" left from the fusion of hydrogen—will itself begin to fuse, forming carbon. Thus the ash of one nuclear fire becomes the fuel for the next.

During this period the sun's outer layer again expands, throwing some of its mass into space in the form of a strong solar wind. At its largest, the sun's surface will extend past the present orbit of Venus. Its surface will be cooler, however, because the same amount of energy will be passing across a much larger surface. Stars that emit their energy through this sort of large, cool surface are called red giants. When our sun enters this stage, it's possible that even though its surface will extend past the present orbit of Venus, its considerable loss of mass and gravitational pull by then will have caused the planets to move outward. Consequently, only Mercury will be swallowed up, while Earth and Venus will survive in orbits farther out than they currently are. This is small consolation for living things, however, since Earth's oceans will have boiled away long before the sun attains its maximum expansion.

Once all the helium and hydrogen atoms in the sun's core have been used up, the sun will again start to collapse, this time shrinking

EMISSION NEBULAE

The Great Nebula of Orion comprises vast clouds of interstellar dust or gas so hot that they emit light. This region of active star birth consists mostly of gaseous hydrogen. Trace concentrations of water molecules also have been discovered here, and astronomers calculate that, due to this nebula's huge size, it holds enough water to fill Earth's oceans 60 times over! Many believe that the water so essential to life on our planet may have originated from just such a stellar source.

down to an object about the size of the Earth—less than one percent of its current diameter. At this point, the electrons in the sun will be so crowded together that they cannot be compressed any more. The contraction will stop and the sun will stabilize. Although no longer generating energy, it will be very hot. A small, hot star like this, one that is no longer undergoing fusion, is called a white dwarf. Think of it as a cooling cinder floating in space for billions of years.

So the stages in the sun's life cycle include its birth enshrouded by dust, its prime as an ordinary main-sequence star, its late fling as a red giant, and its final demise as a white dwarf. When we look at the sky we see many types of stars, many of them representing a star like our own sun at a different stage in its life cycle.

More massive stars—those ten times greater than our sun—have a very different end from the relatively sedate red-giant/white-dwarf pattern. Like the sun, these stars burn hydrogen fuel early in their lives. But they consume their hydrogen much more quickly. Once it is gone, the inevitable collapse occurs, and the helium ash is itself burned to create carbon, as it will be in our sun. But in much heavier stars, the greater force of gravity behind their collapse drives the temperature up to the point where carbon itself is burned. Elements like oxygen, neon, magnesium, and silicon are created by fusion as the star tries to stave off the effects of its own gravity. Eventually, these nuclear reactions begin to produce iron. Iron happens to have the most tightly bound nucleus of any element. It requires energy to break iron into smaller nuclei, and it requires energy to make iron undergo fusion reactions to make heavier nuclei. Iron, in other words, makes a lousy fuel for a star—there is just no way to get energy from it.

Once a star has reached this stage, the unburnable iron ash begins to accumulate in its core. Just as in a white dwarf, electrons become so crowded together that at first they keep the core from collapsing further. Under the inexorable influence of gravity, however, more of the star's material will be burned to make iron, which accumulates until the pressure in the core is high enough to force the electrons to merge with the protons in the iron nuclei. This creates neutrons—particles as heavy as protons, but with no electrical charge. Once this process begins, it rapidly snowballs, and soon the entire mass of the iron core becomes converted into neutrons.

With the electrons gone there is nothing to counteract the effect of gravity, so now the neutron core collapses. In the space of a few minutes the entire center of the star implodes, creating an incredibly dense object about ten miles across—from something that started out much larger than our sun. Titanic shock waves rip through the star's outer envelope, tearing it apart. For a few days the nuclear reactions in the dying star may emit more light than an entire galaxy! This event, the death of a giant star, is what we call a supernova.

In most galaxies a supernova appears about every 30 years. In our own Milky Way, supernovae are not easy to observe, because most of the stars that produce them are located in the plane of the galaxy and thus are shrouded by clouds of dust and gas. We can, however, see supernovae in other galaxies, and they are often truly spectacular.

EVEN MORE IMPORTANT THAN THEIR BEAUTY IS WHAT GOES ON inside them, for all of the heavy chemical elements in the universe are made in supernovae. While elements up to iron are created during the normal lifetime of large stars, elements up to uranium are produced in the extreme temperatures generated by the shock waves in the supernova explosion itself. These elements are blown out into the galaxy, where they eventually mingle with the interstellar dust and gas. We believe that this is precisely what happened to the cloud from which our solar system formed—it was enriched with elements that had been created by supernovae, in stars that had lived and died long before our sun ever began to shine. The calcium in our bones and the iron in our blood, in other words, were made in the cores of other stars, long before our own star began its life.

And the supernova story doesn't end there. While the remnants of the dying star's envelope are being blown into space, strange things can happen in what is left of its collapsed core. If the star isn't too massive, it will continue to collapse until the neutrons can no longer be pushed together. The result is a neutron star, a body only about ten miles across, but as dense as an atomic nucleus. Neutron stars rotate very rapidly, for much the same reason that an ice skater rotates faster when she pulls in her arms during a spin: the diameters of the star and the skater are both reduced. The star will also have an intense magnetic field, because it has pulled so tightly into itself. Particles trapped in its magnetic field will emit radio waves along the direction of the star's north and south magnetic poles. Think of the star, then, as kind of a lighthouse in the sky, emitting a beam of radio waves that sweeps around as the star turns. If we happen to be located in the path of that beam, we will see a bright flash (in the radio spectrum) each time the beam points at us. Radio telescopes on Earth would record this phenomenon as a series of pulses of radio waves.

A rapidly rotating neutron star of this type is called a pulsar. We know of pulsars that rotate as many as a thousand times a second—a very rapid rotation indeed. Because the rate is so regular, and because we have good equipment to measure timing, we can detect many interesting happenings in these stars. We can, for example, observe "starquakes"—events in which part of the outer crust of a star falls a few feet, and, as a result, changes the star's rotation rate slightly.

There is even a move afoot to use some of the faster pulsars to check our atomic clocks, and correct any fluctuations in their time.

The idea is to watch millisecond pulsars over decades and compare their regular, repeated pulsings to our clocks. Some scientists estimate that such pulsar clocks could be as much as a hundred times more accurate than the atomic clocks we use now.

This move is still in its infancy, but it has a pleasant historical ring. After all, the very first standards of time used by humans depended on astronomical events: the rotation of the Earth and the seasonal movement of the sun in the sky—what we call our day and year. It would be fitting if we at some time again turned to the sky to define our units of time, just as our forebears did so long ago.

For very large stars—those with more than 30 times the mass of the sun—the inward pull of gravity continues, causing a collapse that goes past the level of the neutron star, crushing the star into one of the strangest bodies ever conceived—a black hole. Black holes are so small and dense that nothing, not even light, can escape from the immense gravitational pull at their surfaces. Anything falling into a black hole, even the electromagnetic radiation we call light, never comes out. A black hole neither emits nor reflects light, so it is truly black, invisible to our eyes. But that's not to say it doesn't exist. In a sense, it is a system that has so distorted the fabric of space and time around itself that it has cut itself off from the rest of the galaxy. Our galaxy appears to be littered with these so-called stellar black holes, the final stage in the life of very massive stars.

Because black holes don't reflect or emit light, they are very hard to detect. (What would you look for, after all?) The best candidates involve binary star systems in which two stars circle each other but only one becomes a black hole. Since the black hole exerts a powerful gravitational force, we can observe its gravitational effect on its partner star, even if we can't see the black hole itself. So far, about ten candidates for this type of black hole have been found in our galaxy, and more are showing up all the time.

ALTHOUGH WE HAVE CONCENTRATED ON STARS SO FAR, THE Milky Way is more than just a disparate collection of stars. Seen from very far away, it looks like a giant pinwheel rotating slowly in space, with four broad spiral arms bright with stars. Our sun, located about a quarter of the way out along one of these arms, makes a complete circuit of the galaxy every 250 million years.

At the center of the Milky Way is a large globular collection of stars known as the galactic nucleus. The galaxy's central plane—the plane in which it rotates—is full of dust and gas clouds of the type that gave birth to our sun. In some of these clouds, new stars are being born; in other regions, stars are dying—becoming either white dwarfs or supernovae or neutron stars or black holes and spewing heavy elements back into the clouds.

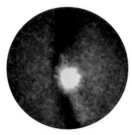

BLACK HOLES

Equipped with gravitational fields so strong that nothing, not even light, can escape them, black holes are incredibly dense yet invisible to us. But as material falls into a black hole, it emits radiation, which we can detect. This composite view from the Hubble Telescope combines visible and ultraviolet images of the area around a particular black hole whose central dust disk seems to be bent like the brim of a hat.

In fact, many nebulae we see in the sky are not distant galaxies of the type studied by Hubble, but relatively nearby clouds of dust and gas. Some represent expanding clouds that are the aftermath of a supernova. Originally, all such fuzzy patches—whether they were due to dust clouds, supernova clouds, or distant galaxies—were lumped under the general term "nebula," a fact that helps explain why they were so controversial before Hubble's work. Now that we know what they are, they're easier to understand.

HOW CAN WE KNOW SO MUCH ABOUT OUR GALAXY WHEN we're swimming in a sea of dust and gas that obscures so much of it? For one thing, although that dust and gas may absorb visible light, it doesn't absorb radio waves or infrared radiation. Consequently, we can "see" the center of our galaxy by looking at it with radio waves. The entire visible spiral structure of the Milky Way is about 100,000 light-years across. Close in, scattered spherically around the central plane of the galaxy, are groupings of stars called globular clusters. They also are roughly spherical in shape, and they contain some of the oldest stars in our galaxy, with ages approaching 12-14 billion years. (Astronomers, by the way, estimate the age of distant stars by measuring how many heavier elements have been manufactured in their nuclear furnaces.)

It was observations of the Milky Way's globular clusters that led American astronomer Harlow Shapley to conclude in 1917 that the sun was not at the center of our galaxy. He noticed that the sky had more globular clusters on one side than the other. Assuming that such clusters occur in a spherical arrangement that is centered on the center of the galaxy itself, he realized that Earth had to be far from the Milky Way's core in order for his observation to be explained. Once again, the Copernican dictum that the Earth is not the nexus of the universe was borne out, this time in a setting infinitely more vast than anything Copernicus could have imagined.

Just outside the Milky Way—about 150 million light-years from Earth—are a couple of small "suburban galaxies" called the Large and Small Magellanic Clouds. Visible only in the southern sky, they honor Ferdinand Magellan, the great explorer and leader of the first European expedition to sail far enough south to see them.

Surrounding both the Milky Way and its globular clusters is a mysterious something we call dark matter. Because it neither emits nor absorbs ordinary light, we can't see it in our telescopes, nor does it emit or absorb radio waves, x-rays, or any other kind of normal radiation. The only reason that we know it exists is that we can observe the gravitational effects it has on visible objects in the universe.

The first inkling that there might be stuff in the universe that we can't see came in the 1950s, when the Swiss-American astronomer

Fritz Zwicky, working at Mount Wilson, noticed that galaxies in some clusters were moving relative to each other (which is normal) at high speeds (which is not). His calculations seemed to indicate that the galaxies were moving so fast that they should all be flying away from each other, not staying together. The only thing holding them in had to be the gravitational attraction of other galaxies, but there wasn't enough visible mass to do that, at the speeds with which the galaxies were moving. Zwicky concluded that there was something in those clusters—some mass capable of exerting a gravitational pull—that he couldn't see. With his work, the question of what we now call dark matter rested for several decades.

In fact, documenting the existence of dark matter proved a very interesting story. In the far reaches of galaxies we can detect individual atoms of hydrogen floating in space. Think of each atom as a microscopic satellite of a galaxy. By looking at the radiation they emit, astronomers can tell how fast they are moving. If there was nothing beyond the visible part of these galaxies except emptiness, you would expect those hydrogen atoms farthest from the center to be moving the slowest—for the same reason that outer planets such as Jupiter and Saturn move more slowly than the inner planets.

During the early 1970s, however, American astronomer Vera Rubin noticed that this wasn't happening. By carefully measuring the faint radiation emitted by hydrogen atoms circling in distant galaxies, she was able to show that atoms at great distances from the center of a given galaxy all move at roughly the same speed. The only way to explain this would be to assume that the atoms are locked in— presumably by the force of gravity—to some unseen material that rotates with the galaxy. Like the stuff that Zwicky thought existed in clusters, this material, whatever it was, couldn't be seen. We knew about it only through its gravitational interactions with visible material. When astronomers estimated the amount of dark matter needed to make those hydrogen atoms behave as they do, they concluded that at least 90 percent of all the matter in galaxies like the Milky Way was in this new, invisible form. Which of course meant that all the stars and dust and nebulae we see, even with the biggest telescopes, account for less than 10 percent of the universe.

THERE ARE MANY GUESSES AS TO WHAT THIS DARK MATTER might be. It may consist of small, non-luminous objects made of ordinary matter—scientists call such things "Jupiters" to conjure an image of a body too small and too distant to be seen. Another possibility is that the region outside the Milky Way's spiral arms is littered with black holes, the massive remains of stars that no longer shine. Still another is that dark matter is made up of neutrinos, subatomic particles created in the early moments of the Big Bang.

Known as 30 Doradus, this emission nebula in the Large Magellanic Cloud is a hotbed of star formation, ionized plasma, and supernova explosions. All these energetic processes make it glow across the electromagnetic spectrum: Red represents x-ray emissions created by gas as hot as a million degrees Celsius, green signifies ionized hydrogen, and blue indicates sources of ultraviolet radiation.

Because 30 Doradus contains some of the hottest, largest, and most massive stars known, experts expect that in a few million years it will give rise to a new globular cluster, a major group of long-lived stars that orbits the core of a galaxy.

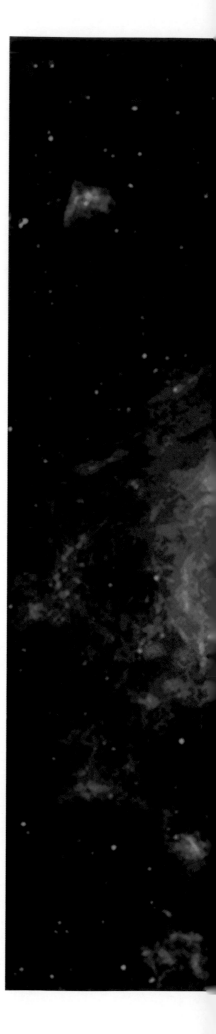

Finally, there is the possibility that dark matter is composed of any one of a number of exotic, theoretical particles that have been predicted to exist but have not yet been seen in the lab. I favor the last possibility. The nature of dark matter remains one of the great unsolved questions of astronomy, but its existence cannot be doubted—we have measured its gravitational effects over and over again.

SCIENTISTS ARE NOW TAKING TWO APPROACHES TO searching for and identifying dark matter. One is an attempt to detect particles of the stuff as Earth moves past it, the other is to see if we can pinpoint concentrations of dark matter within the Milky Way.

In a tunnel excavated deep beneath the Italian Alps, researchers are now monitoring nine 20-pound cubes of sodium iodide, placed there to isolate them from the usual barrage of radiation this planet endures. The idea behind this experiment is simple: If the Milky Way really is full of invisible dark matter, then the sun must sweep though that material as it moves around the galaxy. This means that observers on Earth should be able to detect a "dark-matter wind" blowing by, just as a passenger in a moving car will feel a breeze even on a still day. Occasionally, one of the constituent particles of this dark-matter wind will find its way down the tunnel and interact with a single sodium iodide atom, causing it to send out a tiny pulse of light that can be measured by sensitive detectors. The hope is that, over the course of a year, we will detect differences in the number of flashes—since Earth moves in the same direction as the sun for part of its orbit and in the opposite direction for the other. In fact, researchers announced some tentative results in 1998 that seem to confirm this notion. Once more data has been gathered, they hope not only to provide the first direct evidence for dark matter, but also to say something about the particles that make it up.

A second type of search for dark matter involves looking for small, massive bodies in the spherical outskirts of the Milky Way, in what astronomers call its halo. The idea is to locate one of those bodies as it comes between us and a star in either of the Magellanic clouds. When this happens, the theory of general relativity predicts that light from the star will be bent around the intervening object. In effect, the object acts as a lens, focusing the starlight. To an observer on Earth, the star appears to brighten and then dim again over a period of several days to weeks. This technique has the advantage that any object capable of exerting a gravitational force— a black hole, for example—can be detected by the lensing effect of its gravitational pull, whether it can be seen directly or not.

The search involves fitting telescopes with electronics that enable them to monitor a million stars at a time, waiting for one to show evidence of lensing. Astronomers have seen a number of such events, and are building up an inventory of the small yet massive objects that surround the Milky Way's disk. They are also inventing whimsical names for the objects of their search. One type of particle that might be in the dark-matter wind, for example, is called a WIMP, for weakly interacting massive particle; bodies that produce gravitational lensing have been dubbed MACHOs, for massive compact halo objects.

Imagine what it would be like to approach our Milky Way from the outside. First we would encounter the sphere of dark matter, then the scattered array of globular clusters amid the halo. After that we would enter the more familiar and flatter, spiral-armed part of the Milky Way, scattered with dust and stars. Most of what we know about different types of stars and their life cycles comes from observations made in this visible, relatively nearby neighborhood.

As we move toward the Milky Way's center, the density of stars increases dramatically. At the spherical nucleus, stars pack together so closely that if Earth were located here, we could all read at night just by starlight. At the galaxy's very core, the stars seem to be revolving very rapidly around some central mass. Currently, the best way to explain the motion of these swirling clouds is to conclude that a black hole, perhaps a million times more massive than our sun, lurks here, slowly drawing material into it. Observations of other galaxies seem to indicate that they, too, are centered on massive black holes.

That, in brief, is our home galaxy: a central black hole surrounded by a nucleus of tightly packed stars and spiral arms, all embedded in a gigantic sphere of dark matter. Furthermore, a majority of all the galaxies we see around us seem to be built the same way. But before we leave the Milky Way, let's ask one more question, perhaps the most interesting one of all. Is there anyone out there like us? Among all the billions of suns in our galaxy, is there a planet that has given birth to living things that have developed the kind of technology that would allow us to communicate with them? Or are we alone?

THE QUESTION IS AS OLD AS HUMANITY. AS A LEGITIMATE subject for scientific investigation, however, it is fairly new. It concerns what's often referred to as the Search for Extraterrestrial Intelligence, or SETI, and it can be traced to a small conference that was held at the National Radio Astronomy Observatory in Green Bank, West Virginia, in 1961. At that conference, scientists pointed out that our ability to send and receive radio waves made us eligible for membership in what they called the Galactic Club, an imagined group of advanced civilizations that had formed a communications network throughout the Milky Way.

The goal of the conference was to try to figure out how many members—if any—the club might have, and how we could contact them. The group's approach was straightforward. In order for any advanced civilization to exist, it would need a star for an energy source, a planet capable of developing life, a history that encouraged the development of intelligence, conditions favorable to technological development, and so on. Plugging in the best data available in their day, the scientists calculated that membership in the Galactic Club could be at least several thousand.

Our ideas about the prevalence of life in the solar system have changed dramatically since then, but the notion of a galaxy teeming with intelligent life remains a fixture of our popular intellectual landscape, as *Star Wars* and *Star Trek* testify.

Since 1961, we have learned a great deal about the universe that bears on the question of extraterrestrial life. We understand the importance of liquid water to the development of living things, and realize that only the right-size planet just the right distance from just the right kind of star can have liquid water on its surface for billions of years. We understand that it is possible for life to appear and not develop intelligence—indeed, this would describe the Earth for most of the past four billion years.

We also realize that intelligent life need not develop technology (think of dinosaurs or even whales), and we have no idea whether, or why, a technological civilization would choose to attempt communication. Although most scientists still feel that there must be intelligent life out there somewhere, if only because there are so many stars that could nourish it, few today support the optimistic estimates of the early days of SETI.

My own calculations with astronomer Robert Rood, for example, put the probability that another technological civilization is trying to communicate with us right now at a very modest one in ten thousand. Even so, we both completely support a continued search for communicative civilizations, for in science you can never know how a given experiment will turn out until you actually do it.

THE SEARCH FOR SIGNS OF LIFE BEYOND THE BOUNDARIES of our solar system is continuing in earnest. In fact, the University of California, Berkeley, endowed what must be the first academic chair in the world reserved for the search for extraterrestrial intelligence, a move that underscores SETI's status as an area of serious scientific enquiry. Berkeley also plans to build an array of radio telescopes that will have a collecting area of a hectare—2.5 acres—making it the largest such facility devoted to SETI. In addition, people at the same university have developed a computer screen-saver they call SETI@home, which already harnesses the computational power of hundreds of thousands of Internet-linked personal computers when they are not being used by their owners, by directing them to crunch radio-telescope data. In effect, SETI@home gives ordinary citizens all over the world a chance to participate in science's ongoing search for extraterrestrial life.

A recent and encouraging development is the discovery of planets outside our solar system. By 1999, astronomers had found evidence for about twice as many planets orbiting other stars as the nine known to orbit our own. They relied on the fact that planets don't really circle their stars; rather, both circle a point between them known as the center of mass. (Imagine trying to balance a stick with a star at one end and a planet at the other; your finger will be at the center of mass.) Therefore, a star with a planetary system seems to wobble in space, rather than move in a straight line.

In 1991, astronomer Alexander Wolszczan—using the world's largest radio telescope, in Arecibo, Puerto Rico—announced the first discovery of a planet outside our solar system. Amazingly, the home star of that planet was a pulsar—the remnant of a supernova! Theorists now believe that planets that orbit pulsars originated within a disk of debris, which formed around the pulsar after the supernova event. Obviously, this sort of planet would not be a good candidate for life, due to the intense radiation from the pulsar.

But just a few years later, in 1995, astronomers began to discover evidence for planets circling normal stars. Because the largest and most easily detectable wobble is produced by big planets that are near their stars, we should not be surprised that most new planets found so far are as large or larger in mass than Jupiter and closer to their stars than Earth is to the sun. This caused some consternation at first, because, as we've seen from our own solar system, we expect large planets to form in the frigid regions far from the heat of their stars. By now, however, theorists have become reconciled to the existence of these "hot Jupiters," and argue that they may have formed in dusty regions far from their stars and then moved inward.

Astronomers are also beginning to think seriously about imaging extrasolar planets, in addition to detecting them indirectly. No ground-based telescope can do this, simply because the interference from Earth's atmosphere is too severe. Current plans call for putting rather complex systems of telescopes into space. One suggests launching a collection of satellites, each carrying a small mirror. If the distances between satellites can be kept constant to a high level of accuracy, modern electronics will quickly assemble all of their data and create an image equivalent to the one that would be taken by a telescope whose diameter is the size of the satellite array.

In 2003, in fact, NASA plans a brief mission in which scientists will attempt to control the motion of two side-by-side satellites well enough to use them as a telescope. It won't be easy, for at this sophisticated technological level, even things like vibrations from tape recorders can throw off the navigation. There are also tentative

plans, much farther in the future, for locking satellites together on a grid. Eventually, we can expect such telescopes—or perhaps other, more advanced instruments—to produce images of extrasolar planets and even detect chemical evidence for water and oxygen in their atmospheres. When that happens, the quest for ways of finding Earth-like planets will have succeeded.

LET US NOW LEAVE THE MILKY WAY AND THE MAGELLANIC Clouds to enter what is truly the final frontier—the realm of the galaxies. Looking into space, we see that matter is not scattered uniformly, but instead is collected into galaxies. Most galaxies that we can see with telescopes—about three out of four—are spiral-shaped, like the Milky Way. In them, as in the Milky Way, we see stars in all stages of their life cycles. There are stars being born from clouds of dust and gas, stars thriving, stars dying, and, in some cases, stars ejecting new chemical elements into space. These galaxies are just the sorts of places where we can imagine planets like the Earth circling stars like our sun—and perhaps even producing intelligent beings who, if not human, probably would not seem strange to any reader of science fiction.

Perhaps 10 percent of the known galaxies are elliptical galaxies. Shaped something like giant footballs, they harbor older stars than do spiral galaxies. They range from giants a few million light-years across to dwarfs only a few thousand light-years in size.

If spirals and ellipticals were the only kinds of galaxies, we would conclude that what we see in our own Milky Way is pretty typical of the universe in general. But there are others, including a third major category: Galaxies where titanic explosions and cataclysmic convulsions blow huge clouds of matter across thousands of light-years of space, galaxies which, for reasons that we do not yet understand, put out much more energy than does the Milky Way. These are the so-called active galaxies, places where the universe just seems to go crazy, where energy levels are higher than anywhere else, sometimes vastly higher. Astronomers arrange active galaxies in a hierarchy based on the amounts of energy they emit.

The most common active galaxies are the radio galaxies, the first of which was discovered in 1946. Members of this type tend to emit a lot of radio waves—up to ten million times what a galaxy like the Milky Way emits. They appear fairly mundane when viewed in visible light, except that they often have lobes pushing out from their sides, as though the galaxy were expelling clouds of matter in jets or beams from its very center into intergalactic space. We know of tens of thousands of radio galaxies and are discovering new ones daily.

Perhaps the most mysterious objects in the galactic sky are quasars, which we now believe are extremely energetic centers of active

GALAXIES

Jeweled wheels of the universe, galaxies each contain thousands of star clusters, billions of stars, and enormous amounts of diffuse gas, all orbiting a central point.

Astronomers classify them according to basic shapes and activity levels. Here, spiral galaxy ESO 510-13 appears edge-on, revealing a pronounced curve, or warp, to the plane of its disk, which may indicate past interactions or even collisions with other galaxies. Our own galaxy, the Milky Way, is believed to have a small warp. About 150 million light-years from Earth, ESO 510-13 spans some 100,000 light-years in diameter.

galaxies. The first quasar was discovered in the 1960s, when astronomers determined the precise position of an unusual cosmic radio source and gave it to astronomer Maarten Schmidt of Caltech. Schmidt determined that the object had to be several billion light-years away from Earth, making it the most distant object by far that had ever been seen. The fact that we can detect quasars at all, given their enormous distances from us, means that they are pouring immense amounts of energy into space.

Scanning the night sky is rather like approaching an unfamiliar city in an airplane: the first things you see are the brightest lights in town. In the same way, astronomers studying the farthest sectors of the universe focus on the most energetic formations there. Quasars remain some of the most distant objects known, although pinpointing exactly which quasar is farthest changes from month to month, as astronomers get better at detecting more distant (and fainter) ones.

The puzzle of quasars is deepened by the observation that some of them have flare-ups in which their brightness increases markedly for a period of days or weeks. This means that whatever their energy source is, it must be fairly small—no more than a few light-days or light-weeks across. How do they generate so much energy in so small a space? We don't know. It's a little bit like finding a new fuel that can run your car for a month on a single chip the size of your fingernail.

CURRENTLY, WE BELIEVE THAT ALL ACTIVE GALAXIES, FROM radio galaxies to quasars, result from the action of very massive black holes. I don't mean relatively puny black holes like the one at the center of the Milky Way, which has a gravitational effect equivalent to a mere million suns or so. I am talking about black holes whose masses are billions of times more massive than our sun. They are truly ravenous, sucking all kinds of material into them.

The erratic nature of active galaxies, the explosions and eruptions we witness going on in them, might logically be the result of clumps of material heating up and colliding as they fall into these black holes. Some astronomers have even suggested that the most energetic active galaxies—quasars—require not just the matter from a single galaxy to fuel them, but are the end result of a collision between galaxies. If this is the case, the energy involved amounts to millions of stars. Like all black holes, those of active galaxies are not visible in themselves. But they can be detected through their effects on the matter falling into them.

Even so, the universe we see beyond the Milky Way is a diverse one, with all sorts of strange objects, whose various arrangements and movements are not at all what we might have first expected.

To understand what I mean, let's return to Edwin Hubble on his California mountaintop in 1925. In addition to establishing the existence of other galaxies, Hubble noted that when he looked at the light emitted by atoms in those galaxies and compared it to the light emitted by similar atoms in terrestrial laboratories, the galactic light always appeared redder—that is, it had a slightly longer wavelength. The signature series of colors that characterize each atomic element was there, but they were all shifted equally. Hubble interpreted this shift as being due to the motion of the galaxies from which the light originated. In fact, this "Hubble redshift" is an example of a phenomenon known as the Doppler effect.

We tend to be more familiar with the Doppler effect for sound than for light, but the principle is the same in both cases. If you're standing on a highway and a car blows its horn as it goes by, you hear a change in pitch as the car passes. This is because as the car approaches, the crests of the sound wave it emits are more tightly packed together than when it is standing still, and your ear interprets this compacted wave as a higher pitch. In the same way, when the car passes you and is moving away, the wave crests are farther apart— causing you to hear a lower pitch.

Substitute light for sound and color for pitch, and you understand Hubble's conclusions. If light from distant galaxies is consistently shifted toward the red end of the spectrum, that is, toward longer distances between crests, then those galaxies must be moving away from us. Furthermore, the larger that shift, the faster the source galaxy must be receding from Earth.

When Hubble compared his redshift data to known distances of various galaxies—measured by using the Cepheid-variable standard candle—he found a clear regularity. The farther away a galaxy was from Earth, the faster it was moving. This result, based on a small sample of galaxies in Hubble's time, has been borne out repeatedly in subsequent measurements, and is now known as Hubble's Law. Its inescapable implication is that the universe is expanding.

HERE'S AN OFT-USED ANALOGY THAT SHOULD HELP YOU visualize what the Hubble expansion is all about: Imagine a lump of dough of raisin bread that's rising in a warm kitchen. As the dough continues to rise, the raisins move farther and farther apart, because the dough between them is expanding. The farther each raisin gets from the next, the faster it will move away from the other, because there's more dough between them that can expand. To someone standing on a raisin, it will appear that the entire chunk of dough is expanding regularly, and what Hubble's measurements teach us is that our universe actually is expanding just this way.

This picture of the Hubble expansion explains several things about our universe. First, it explains why Earth seems to be at the center of the expansion, for any raisin in the dough will see itself as the center,

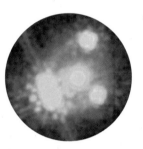

QUASARS

Discovered in the early 1960s, quasars rank among the most distant objects in the universe, as well as the brightest. Some radiate as much energy as a thousand Milky Ways, yet take up little more space than our solar system. The four blobs of light visible here stem from the same quasar, but a galaxy (orange glow) lying between it and Earth acts like a lens, its gravitation bending the quasar's yellow light into four images. This quasar lies eight billion light-years away, in the constellation Leo.

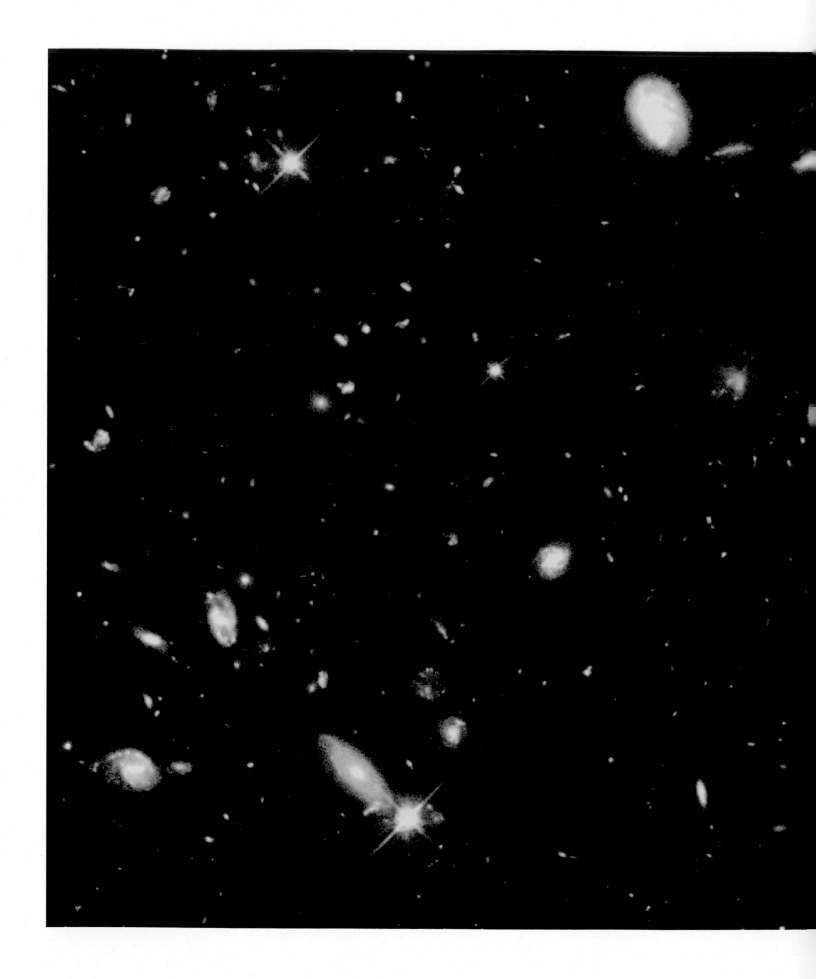

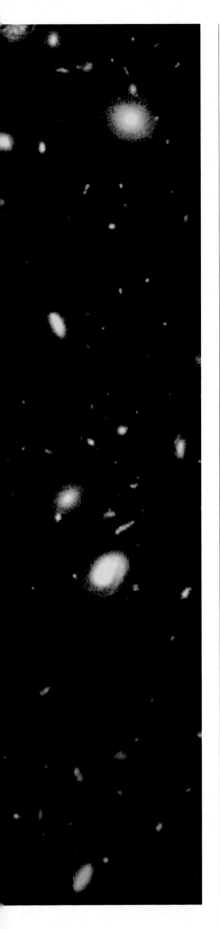

since everything else is moving away from it. Second, it shows that the universe must have begun at a specific point in the past, since it is expanding regularly. Imagine running the film backward on the Hubble expansion, and you come up with a universe that began in a hot, dense state perhaps 12 billion years ago and has been expanding and cooling ever since. Both the initial event—whatever it was that initiated the universe—and the subsequent cooling and expansion are included in the term Big Bang.

THE FACT THAT THE REDSHIFT OF EACH GALAXY IS RELATED to its distance leads us to another important idea. Like the stars, galaxies present themselves in a two-dimensional array, giving us no direct way to measure their distance from us. For relatively nearby galaxies, we can use Cepheid variables to determine distances, but we can't do that for galaxies very far away—our instruments simply aren't good enough to distinguish individual stars. By measuring the redshift of an entire galaxy, however, we can use Hubble's Law to convert our two-dimensional picture of the sky to three dimensions. The larger the redshift, the greater the distance to the galaxy.

Astronomers who measure redshifts to a large number of galaxies have discovered that galaxies are not scattered uniformly through space, but seem to be gathered together in huge superclusters. Between these superclusters are regions with few or no galaxies at all. Some people started calling the empty regions Hubble's bubbles, but this has been replaced by a more sedate term: voids.

The universe that has been uncovered by redshift surveys is like a mound of soapsuds in a sink; there are places containing matter (that is, the soap film itself) and places that are empty (the interiors of the bubbles). Our universe seems to have arranged itself very similarly, with broad but thin concentrations of matter surrounding vast regions of emptiness. Redshift surveys remain a major frontier of modern cosmology. In the words of Margaret Geller, with the Harvard-Smithsonian Center for Astrophysics and a pioneer of redshift surveys, "If technology continues to advance at its present pace, we will have mapped out the entire visible universe by the end of the next century. Considering that it took many centuries to map the Earth, this will be an enormous achievement for humanity."

But even when we have such a map, questions will remain. Why should galaxies be arranged in a cluster-and-void pattern? For that matter, why should galaxies form at all? Though the answers continue to elude us, we shouldn't be discouraged or even too surprised. One of the great attractions of science and especially of astronomy is that there will always be unanswered questions and unsolved mysteries. What better way to conclude our tour of the great beyond than to look at some of the areas that engage astronomers at this very moment? ●

DEEP SPACE

Not stars but entire galaxies dot this Hubble Telescope image, which takes in one of the sky's emptiest regions—an area roughly the size of a sand grain held at arm's length. (Only a few of the points of light visible here are individual stars.)

This is our universe: layer upon layer of other worlds, as far as the eye—or the Hubble Telescope—can see. Light from some of the galaxies in this view has taken 11 billion years to reach us. The large white blob at top center is one of the nearest, hovering a mere four billion light-years away.

of the columns in a process called photoevaporation, creating eerie streamers and highlighting their three-dimensionality.

This trio composes but a tiny part of the Eagle Nebula, also known as M16, which lies about 7,000 light-years away in the constellation Serpens. The leftmost pillar spans about a light-year from base to tip. Nebulae—Latin for clouds—simply refers to huge masses of gas and dust that, like M16, may generate new stars.

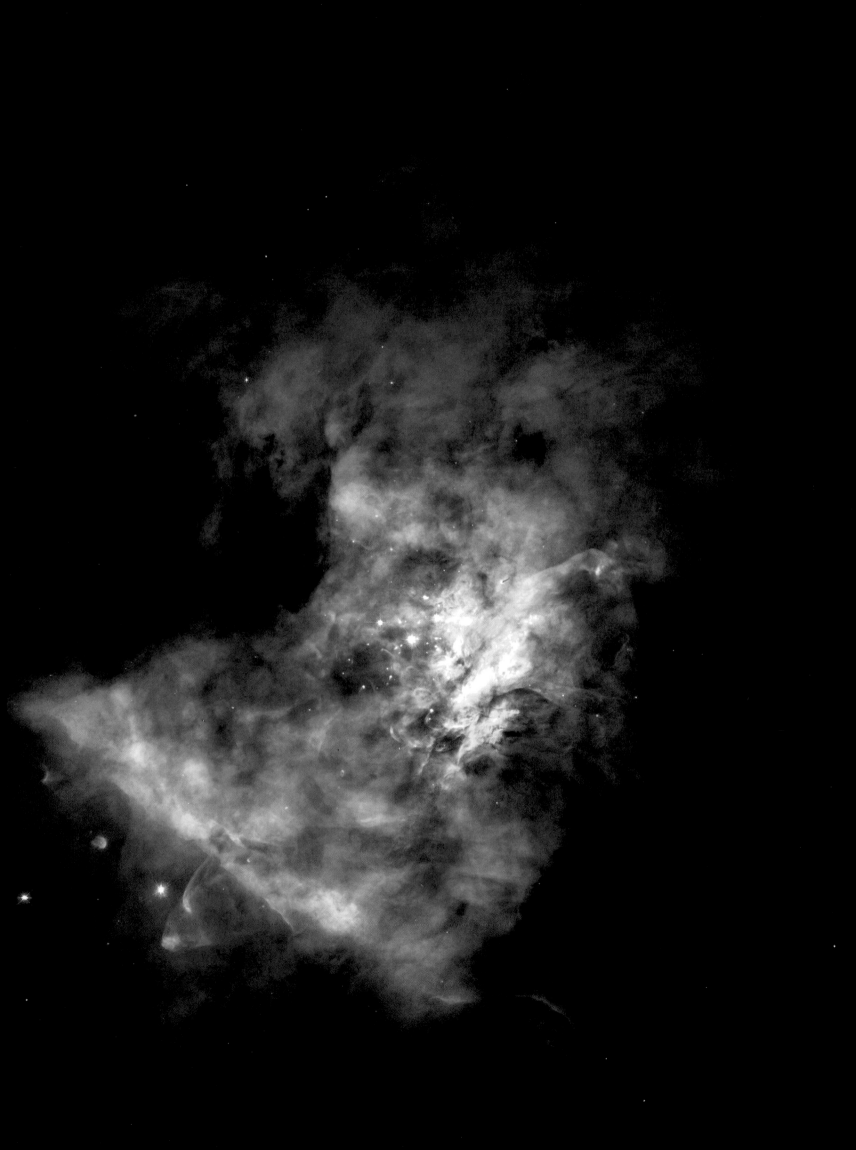

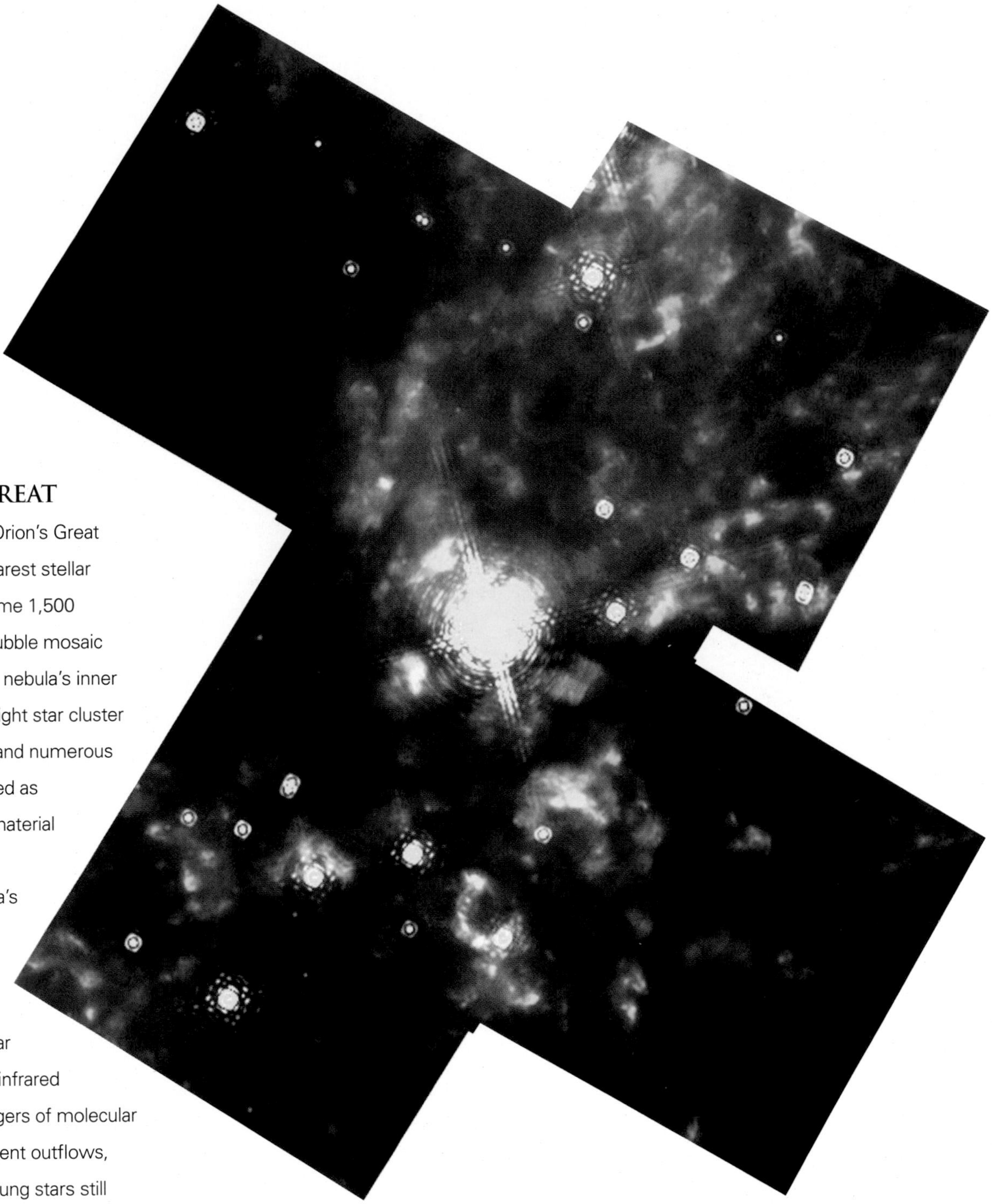

ORION THE GREAT

Swirling vastness of Orion's Great
Nebula boasts the nearest stellar
nurseries to Earth, some 1,500
light-years away. A Hubble mosaic
(opposite) depicts the nebula's inner
portion, including a bright star cluster
called the Trapezium and numerous
shock waves, spawned as
high-speed flows of material
meet slower ones.

At the Great Nebula's
heart (right), newly
born stars and
glowing clouds of
interstellar dust appear
yellow-orange in this infrared
image, while blue fingers of molecular
hydrogen identify violent outflows,
probably driven by young stars still
embedded in the dust.

HORSEHEAD NEBULA

[FOLLOWING PAGES] Panoramic view
of this well-known nebula—one part of
Orion's giant cloud complex—reveals
its trademark feature as a mere tendril
of a much larger opaque dust cloud,
poised against hydrogen gas that glows
red from the energy of hot, nearby stars.
The brilliant white star left of center is
one of the trio that make up Orion's belt.

HELIX NEBULA

Nearest planetary nebula to Earth, the ethereal Helix (above) glows a mere 450 light-years away. Its red-violet rings indicate the presence of nitrogen and hydrogen, energized by ultraviolet radiation from the nebula's central star. A Hubble close-up of the Helix's gaseous envelope (right) reveals thousands of "cometary knots"—named for their heads and tails, the latter always pointing away from their core star. These droplet-like condensations may result when hot, fast shells of nebular gas overrun cooler, slower, previously ejected ones. They are enormous: Each head spans several billion miles across, while the tails stretch over 100 billion miles, more than 1,000 times our distance from the sun.

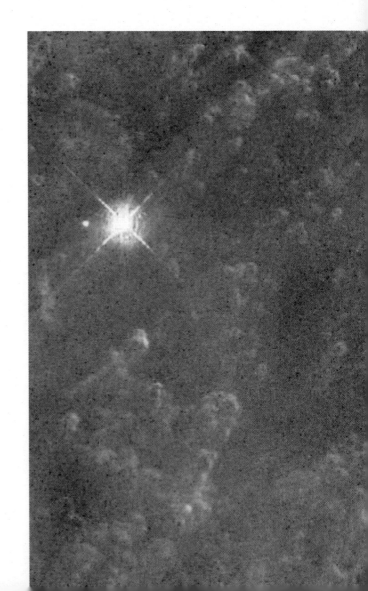

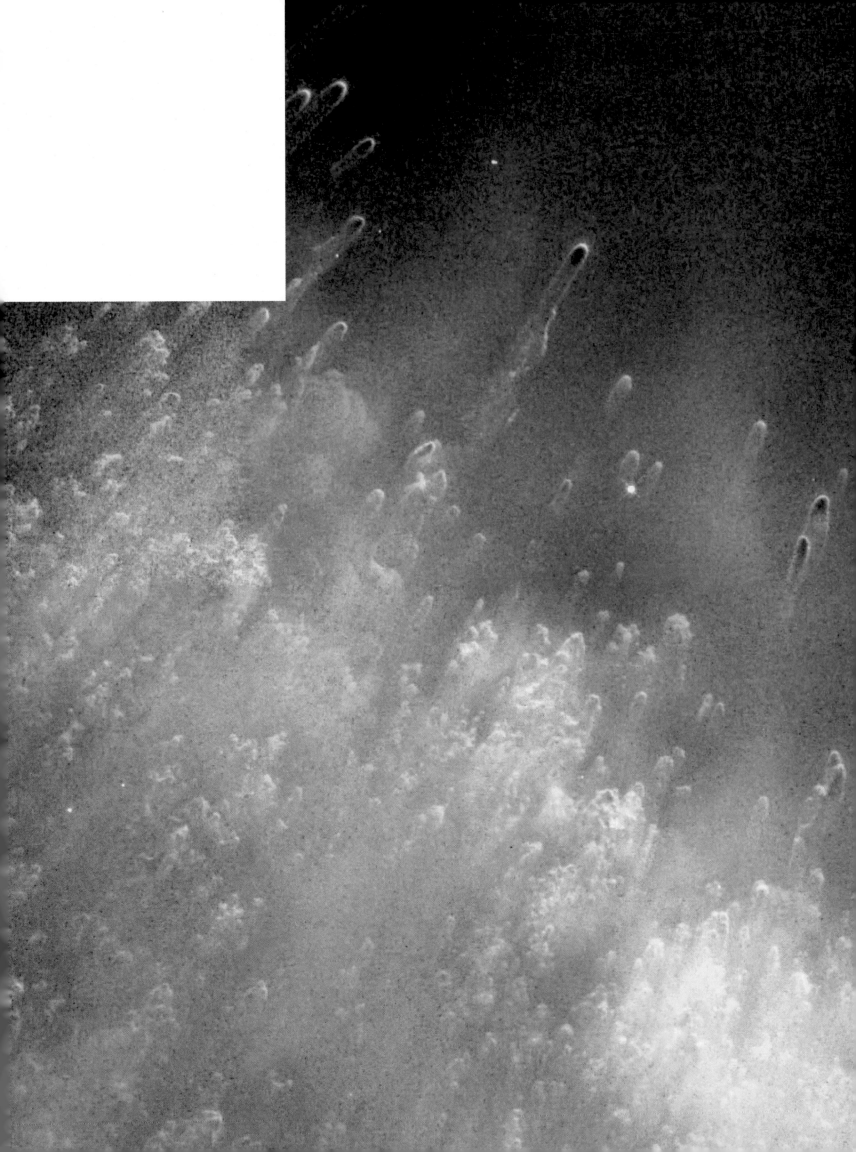

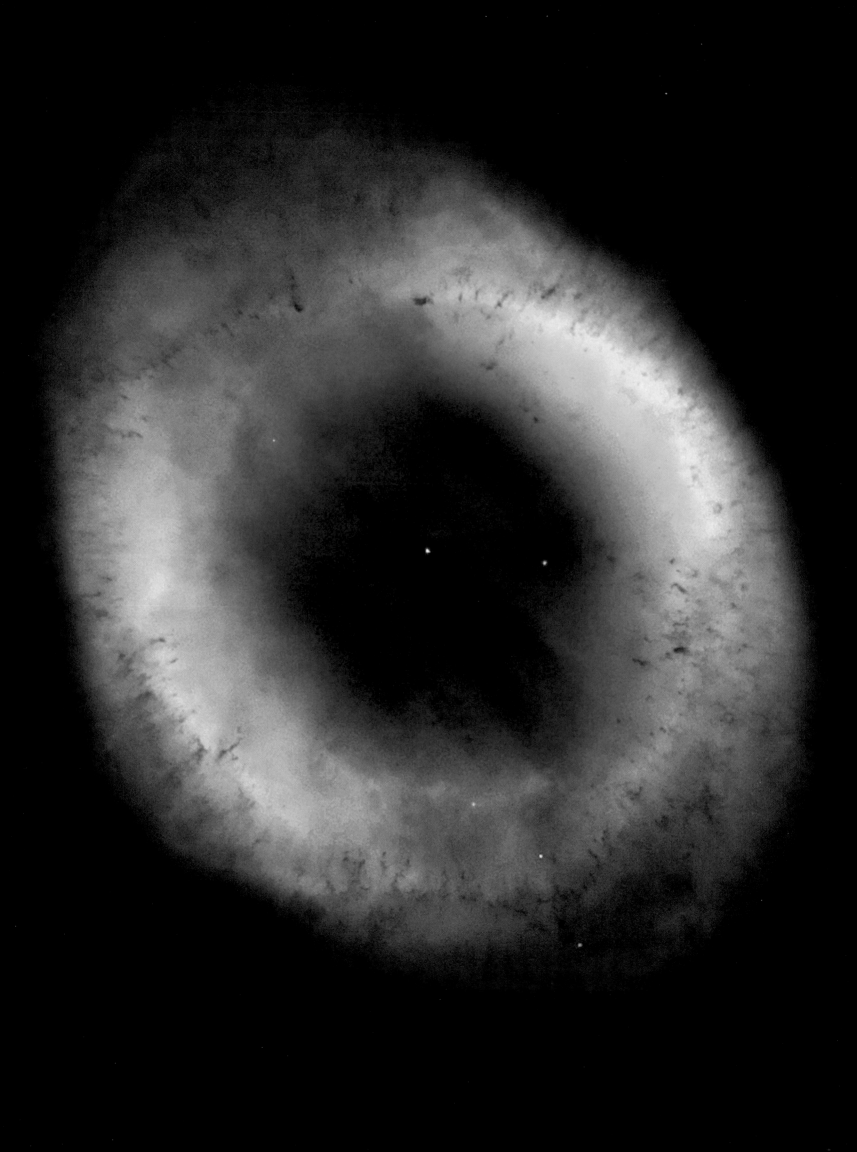

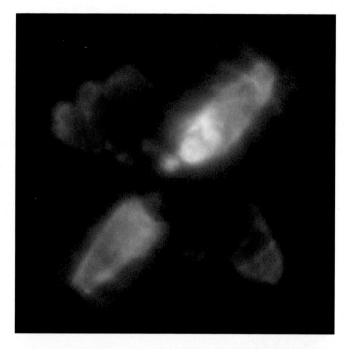

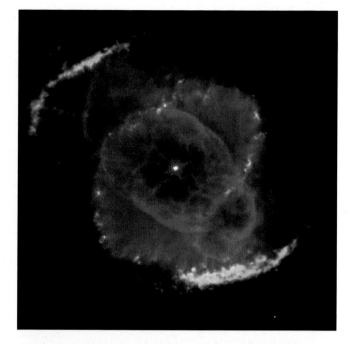

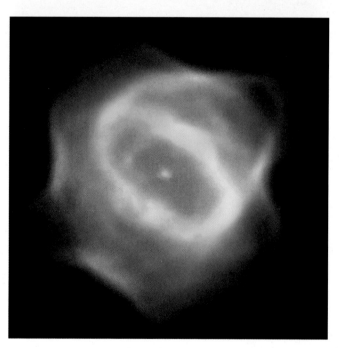

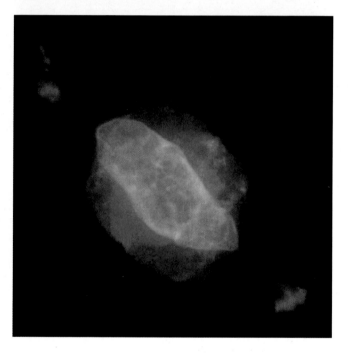

GALLERY OF DYING STARS

Infinitely varied and often whimsically named, planetary nebulae make fascinating studies. The elegant Ring (opposite), in the constellation Lyra, ranks as the brightest. It appears end-on to us, but actually is tubular in shape. Hot blue gas near its central star gives way to progressively cooler greens, yellows, and reds.

Infrared imaging of the Egg (top left) shows multiple emissions of gas amid nebular dust, while the Stingray (bottom left)—the youngest known planetary nebula—billows gas clouds and shock waves. A second star, clearly visible in the core, may account for the sinuous shape.

As complex as it is gorgeous, the Cat's Eye (top right) may also harbor a binary star within its convoluted shells of glowing gas. The highly bipolar Saturn Nebula (bottom right) demonstrates how ejected gas and stellar winds can interact to create jet-like effects.

JET-POWERED BUTTERFLY

Floating like its namesake, the planetary nebula called the Butterfly embraces twin jets of gas with purplish wings. The gas speeds outward at more than 200 miles per second (720,000 miles per hour)! Its core star, one of a pair that orbit perilously near each other, may be engulfing its partner.

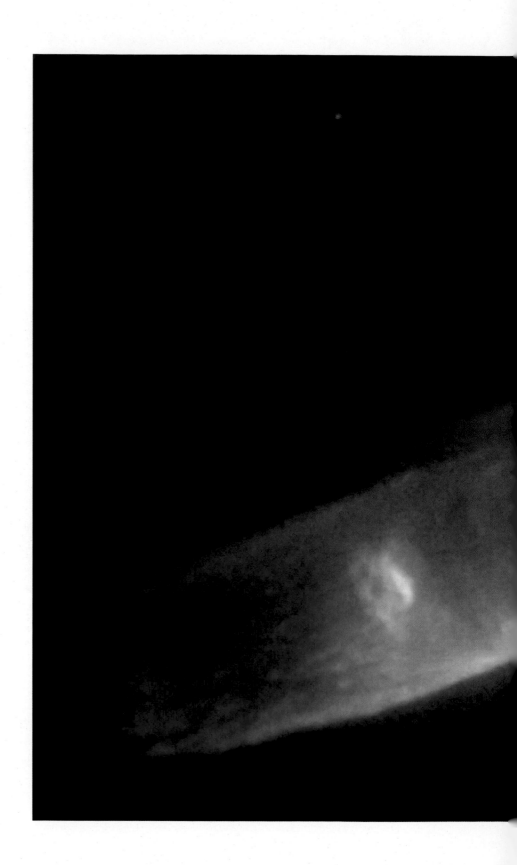

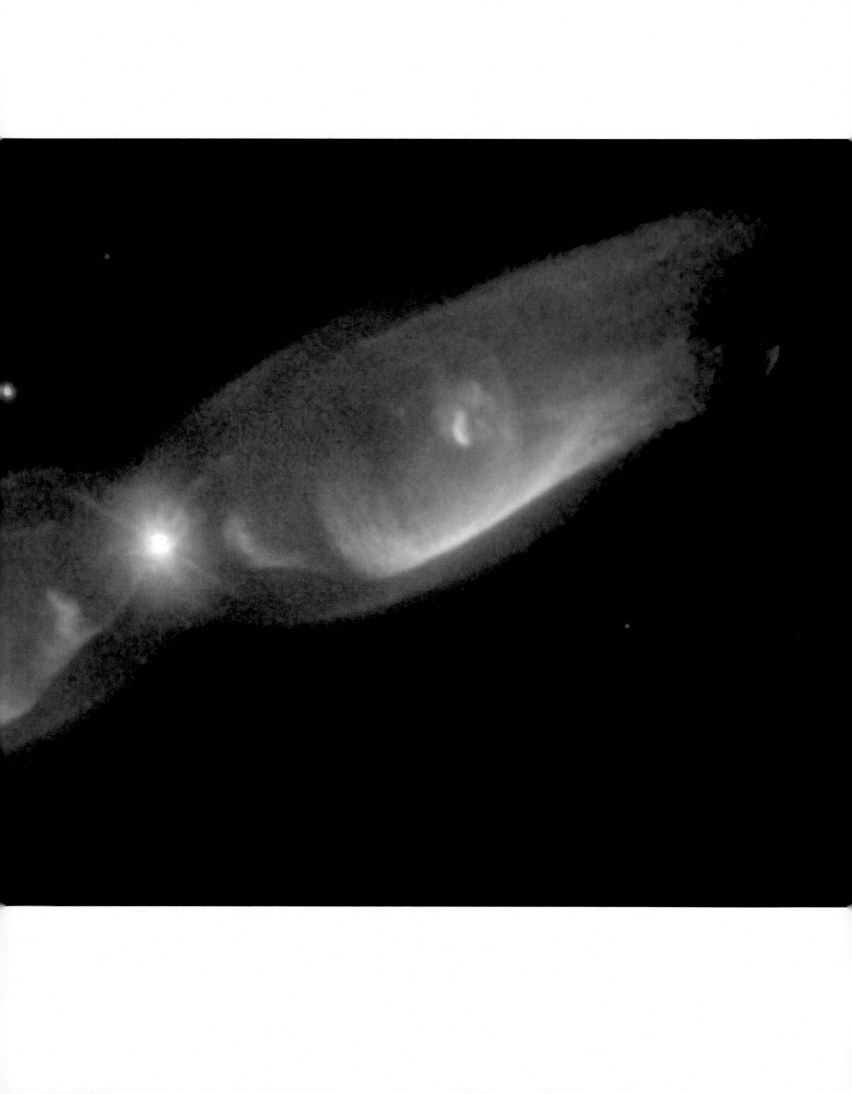

IRRESISTIBLE FORCE, IMMOVABLE OBJECT

Aptly named Bubble Nebula (left) took shape as raging stellar winds from its huge central star pushed a bubble-like shell of expanding gas ever outward. But a nearby molecular cloud is massive enough to resist the expansion. The result?

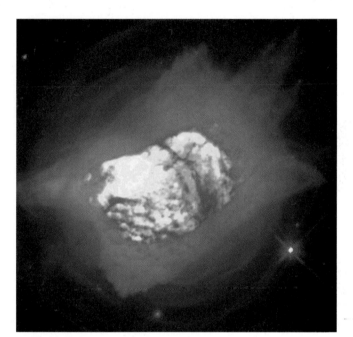

The cloud manages to contain the bubble—for now—but it gets blasted by hot radiations from its core star, causing it to glow. About ten light-years wide, the Bubble can be seen with a small telescope, near the constellation Cassiopeia.

Planetary nebula NGC 7027 (above) shimmers as dense clumps of ejecta produce glowing dust clouds that mask a tiny white dot, the nebula's dying white dwarf star. Our sun can expect a similar end, in about six billion years.

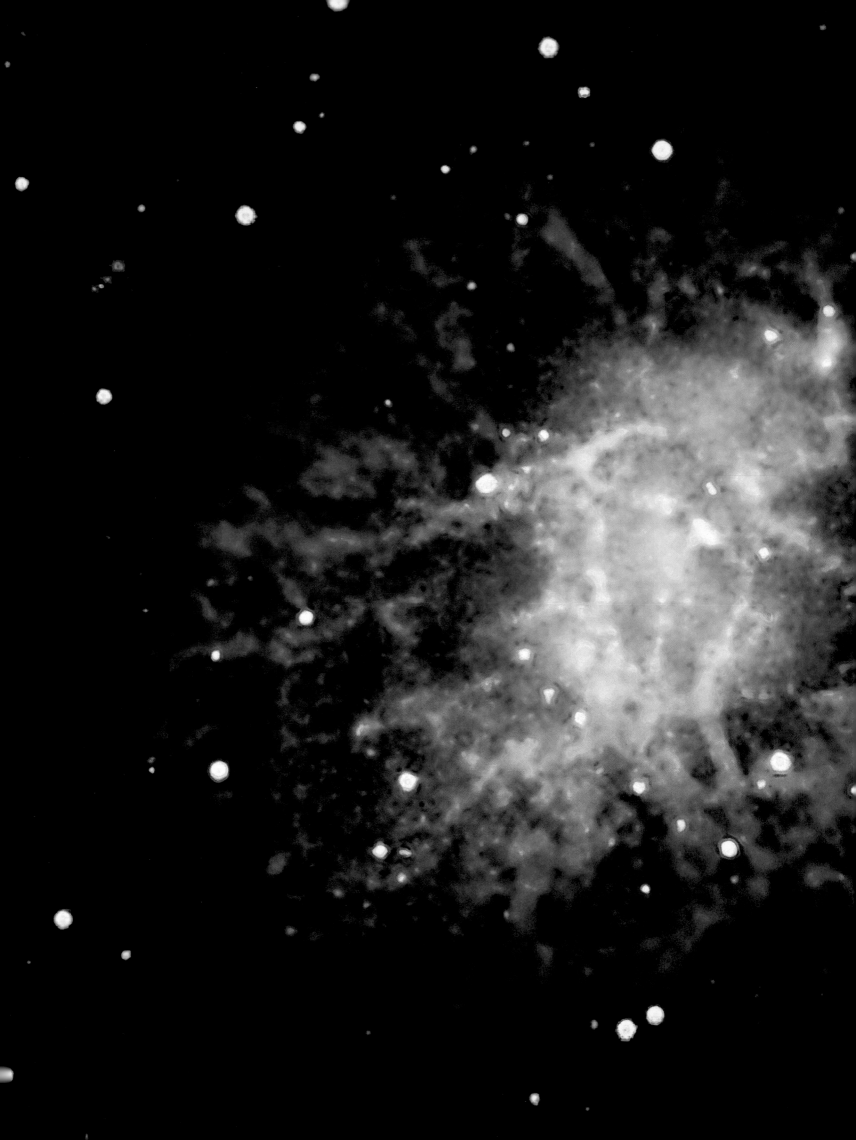

SUPERNOVAE

In 1054, Chinese observers reported a star that became brighter than Venus and then faded slowly over the year. They witnessed what we call a supernova, the colossal explosion of a star too massive to go the red giant/white dwarf route.

Today, lingering remains of that long ago explosion compose the Crab Nebula—the filamentous glob shown here—so energetic that it glows in every kind of known light. At its core lies a pulsar, a rotating neutron star more massive than our sun—but only six miles wide. It expels a beam of particles and radiation that sweeps past us 30 times a second. Red indicates areas where electrons and protons are combining to form hydrogen, while green signifies clouds of free electrons whirling in the pulsar's magnetic field.

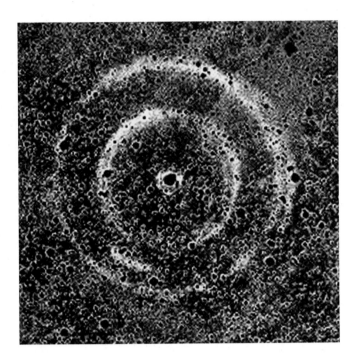

SUPERNOVA 1987A

Echoes of a supernova reverberate
in the Large Magellanic Cloud (left)
where unusual rings and glowing
gas signal the remains of a massive
stellar explosion seen in May 1987.
The rings should brighten in the next
few years, as expanding debris over-
takes previously expelled material,
heating it and causing it to glow. A
view of 1987A's "light echo" (above),
produced from before-and-after
images of the supernova, captures
the true color of its initial blast.

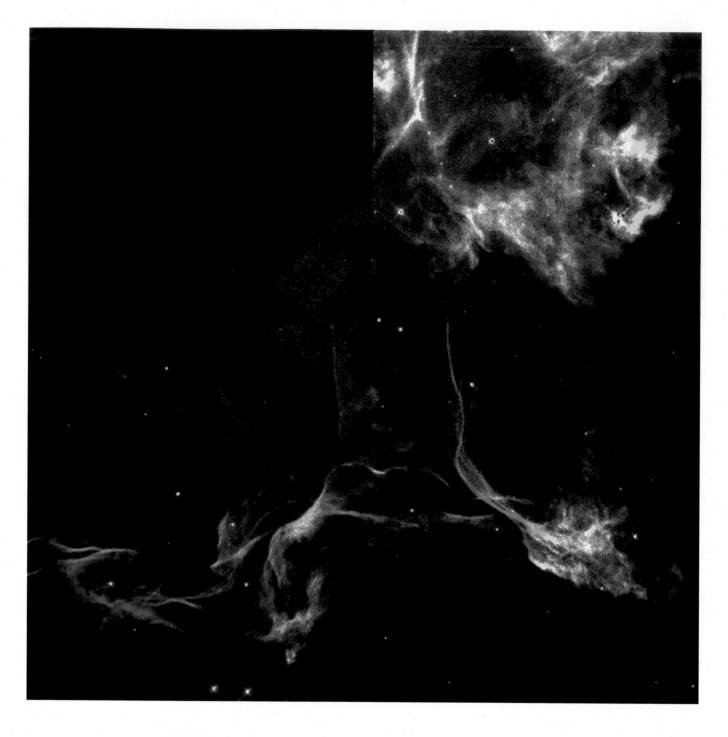

VEILED SWAN

Two views of the diaphanous Veil Nebula, in the constellation Cygnus (the Swan), reveal the continuing effects of a supernova that exploded about 15,000 years ago. A blast wave (above) collides with interstellar gas, sparking its ethereal glow. Blue indicates highly ionized oxygen, red means ionized sulfur, and green signifies hydrogen atoms.

A detailed close-up (opposite) shows that hot gas (blue) catapulted by the explosion has yet to catch up with an earlier gas cloud (red).

GLOBULAR CLUSTERS

Biggest ball of stars in our galaxy, globular cluster Omega Centauri contains
about 10 million separate stars as it orbits the center of the Milky Way.
It is one of about 200 known clusters in the galaxy, which are believed to
have formed before most of the galactic material settled into a disk.

Component stars of each cluster share a common history; many seem to
have evolved past the current stage of our sun, no longer fusing hydrogen
into helium but rather helium into carbon. Soon they shed their outer
envelopes to become the smoldering carbon embers we call white dwarfs.
Their extreme longevity helps establish the age of the universe; studying
them helps explain the processes of galaxy formation.

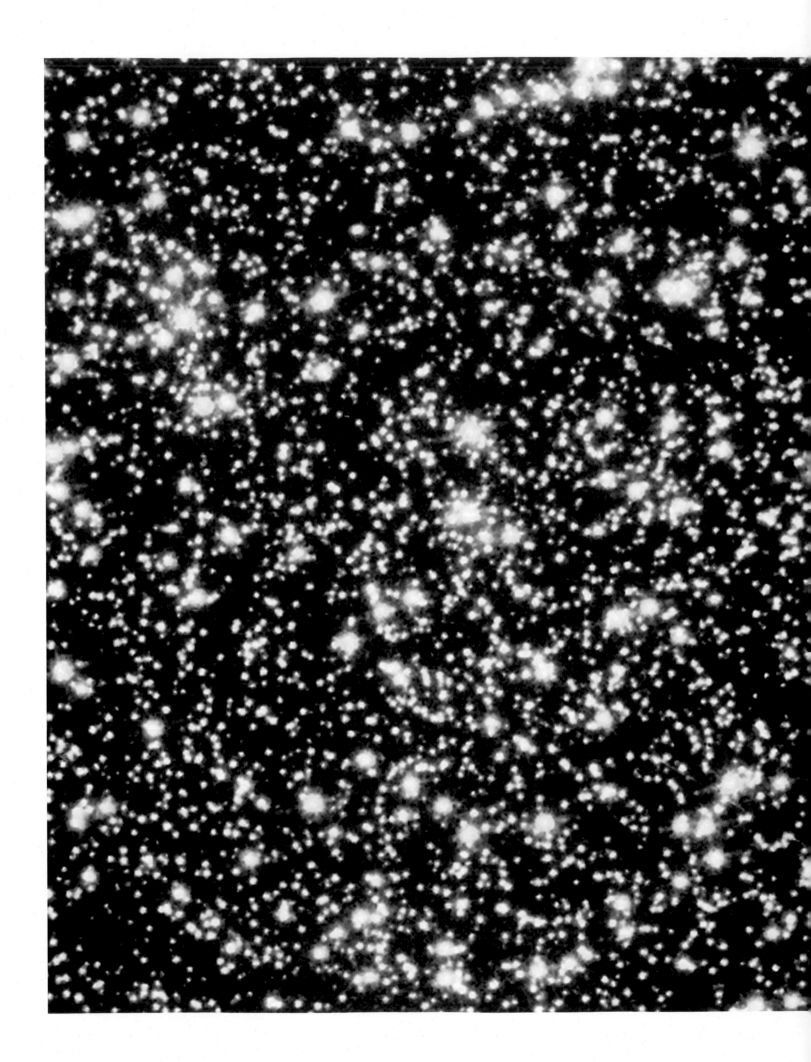

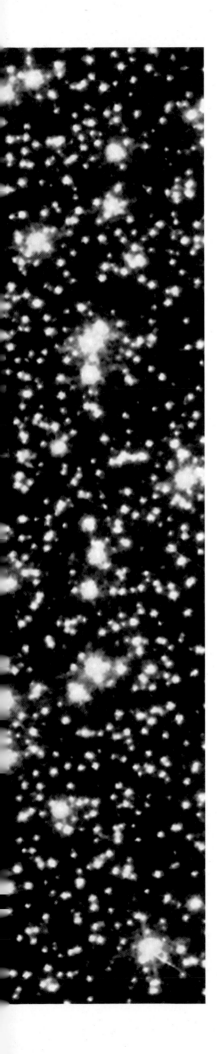

STARS UPON STARS

Peering into the heart of globular cluster 47 Tucanae (left), the Hubble Telescope discovered not only many old stars, but also blue stragglers, which glow with the blue light of younger stars—and may result from the merging of older stars.

Globular cluster G1 (above), in the relatively nearby Andromeda Galaxy, also contains helium-burning stars, whose temperature and brightness suggest that G1 is about as old as our globular clusters, which arose just after the dawn of the universe.

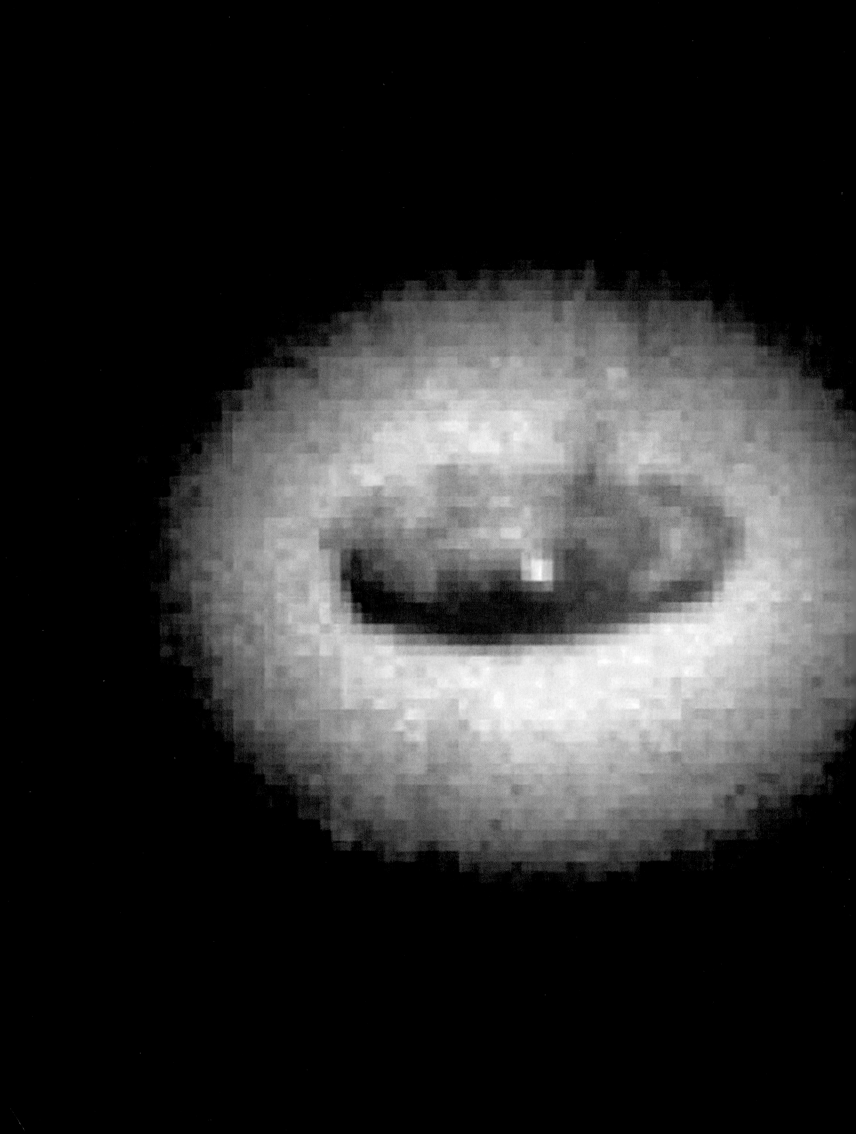

BLACK HOLES & QUASARS

Twin mysteries of the cosmos, black holes and quasars may be different manifestations of the same thing. This swirling disk of gas and dust some 800 light-years in diameter marks what is almost certainly a black hole. It lies near the center of a nearby elliptical galaxy; the disk may be a remnant of a smaller galaxy swallowed by the hole many millions of years ago. Black holes have been found lurking at the hearts of several galaxies, and may be a necessary component of galactic systems.

Quasars, short for quasi-stellar objects, are the most powerful radiators of energy in the universe, the brilliant centers of active galaxies. They may also occur in galactic cores with especially massive black holes. Because all known quasars are very distant, they could hearken back to the dawn of the universe. They may have been more widespread long ago, dying out as their galaxies ran out of fuel to power them.

COSMIC BLOWTORCH

Seething jet of electrons and positrons (left) moves at about the speed of light as it jabs out from a rapidly rotating gas disk at the center of giant elliptical galaxy M87. Studies of the gas disk yield a core density so high that the only thing that could exist there is a black hole.

Signature of another supermassive black hole was found near the center of galaxy M84 (below). Colors reflect rotational velocity, and the pronounced S-shape indicates dramatic swings in orbital speeds of gas already in the black hole's grip, near the core of the galaxy. If no black hole was present, the line would be nearly vertical.

 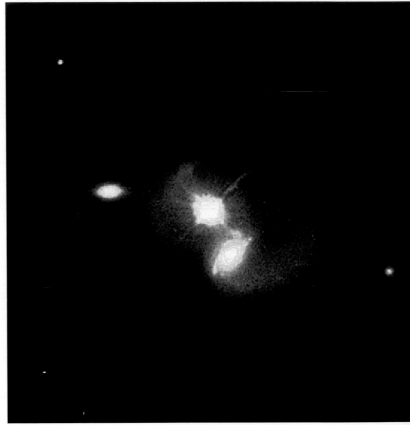

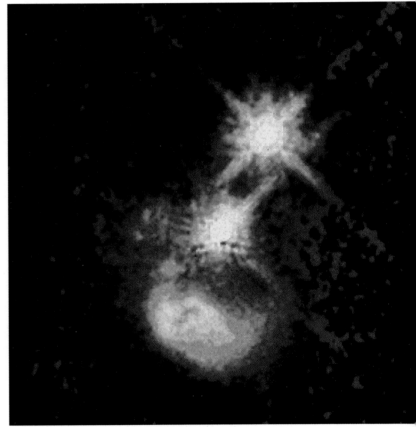

QUASAR GALLERY

Lying near the edge of the observable universe, quasars are so powerful that they can be seen billions of light-years away. Many astronomers believe the quasar's central engine is a gigantic black hole, fueled by enormous amounts of in-falling gas, dust, and stars.

Of the four different quasars here, one (far left) occurs at the center of an apparently normal elliptical galaxy. The others, displaying obvious diffraction spikes, occur in disrupted galaxies that are colliding or merging with neighboring galaxies. Their immense gravitational fields interact, wreaking havoc—and offering plenty of debris to feed the hungry black holes that are thought to dwell at the cores of quasars.

GALAXIES

Island universe awhirl in the blackness of space, galaxy M100 is one of the brightest in the night sky, despite being tens of millions of light-years distant, in the Virgo cluster of galaxies. Like our own Milky Way, M100 is a spiral galaxy, a disk-like affair with a central bulge that contains older stars and globular clusters. Its curving arms mark areas where new stars are forming. Fortunately for stargazers, M100 is oriented near-perfectly face-on to Earth.

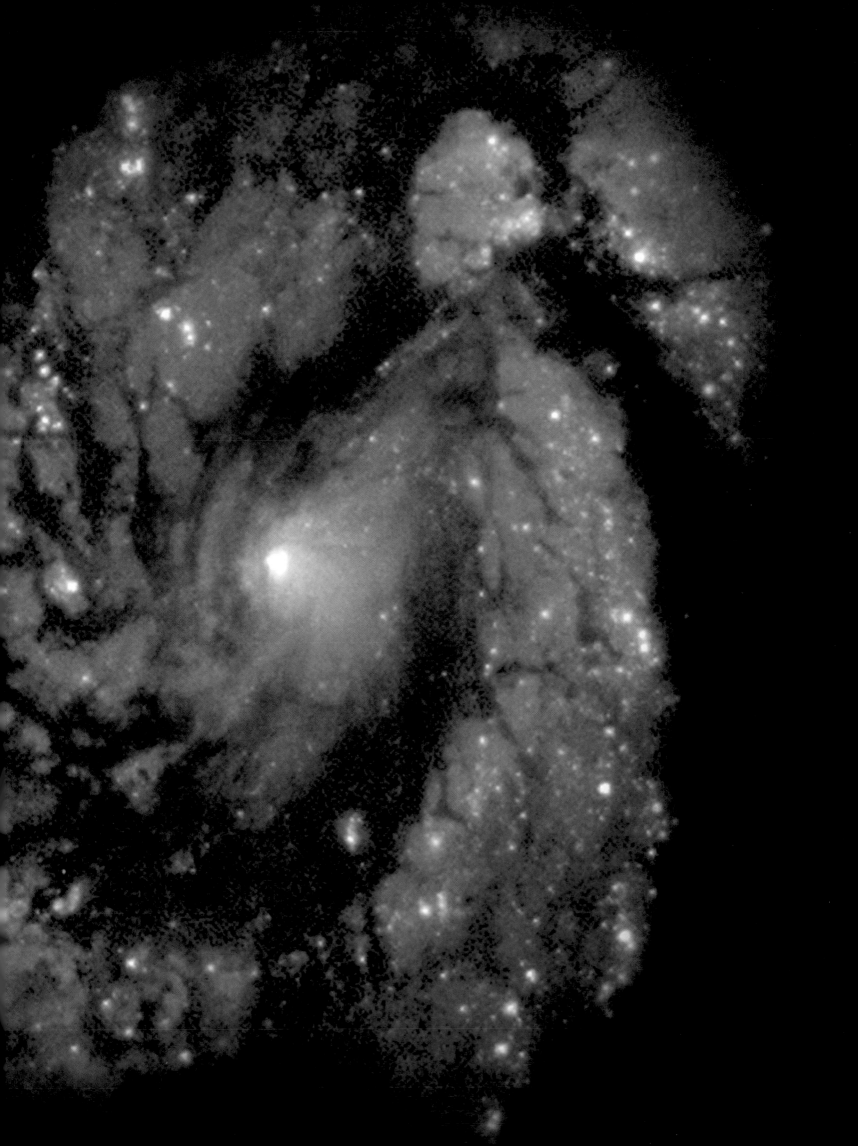

EDGE-ON VIEWS

Two spiral galaxies reveal variety even among similarly shaped galaxies. NGC 891 (opposite) packs most of its gas, dust, and stars in a relatively thin disk. Its dark rim is a lane of light-absorbing dust. Once thought to be similar to the Milky Way, NGC 891 is now known to display unusual filamentary patterns, perhaps indicating that exploding supernovae threw dust out of the disk.

Spiral galaxy M104—commonly known as the Sombrero (above)— also features a prominent dust lane as well as a substantial halo of stars and globular clusters. Its strong x-ray emissions and the high orbital speeds of its central stars may indicate that a black hole equal to a billion suns lurks at Sombrero's center.

COSMIC MERGER

[FOLLOWING PAGES] Gorgeous sprawl of the Antennae galaxies shows two colliding spirals distorting each other as their interacting gravitational fields spawn huge gas clouds and a firestorm of star birth. The cores of these galaxies remain separate; surrounding dust casts them in blazing yellow and red. New globular clusters—many of them blue, indicating massive, hot, young stars—billow along a wide band of chaotic dust. Reddish knots could be clusters that have not yet expelled dust from their systems.

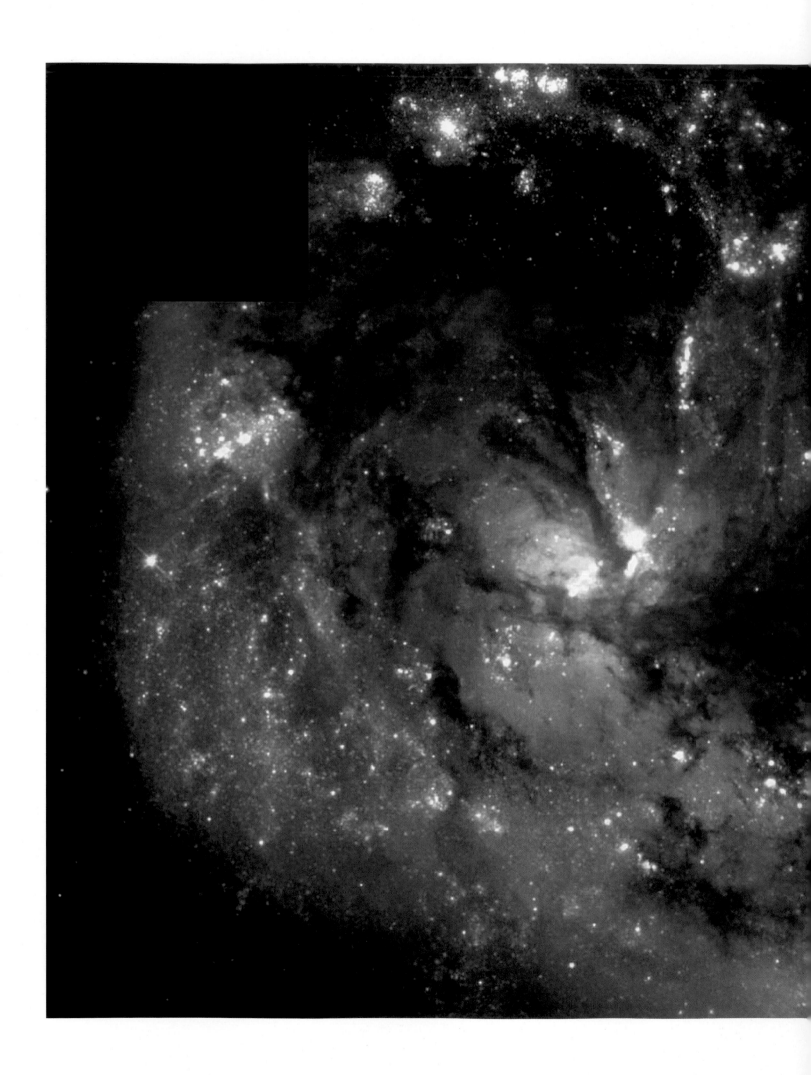

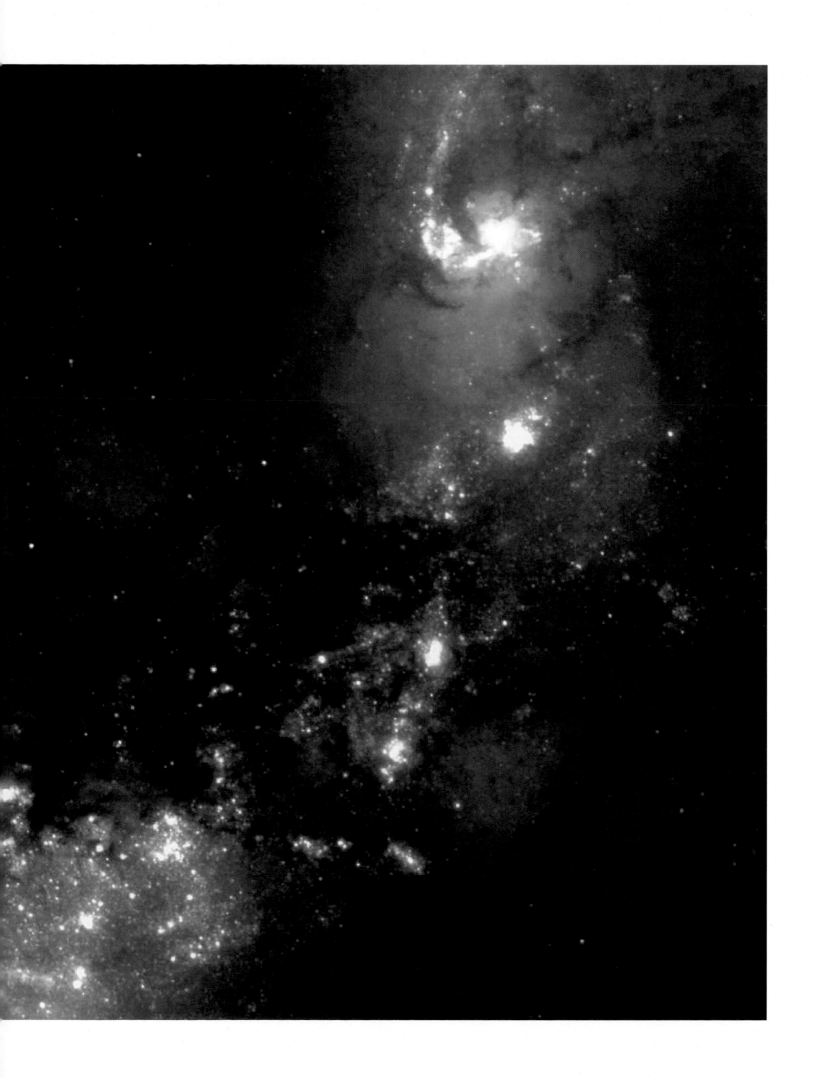

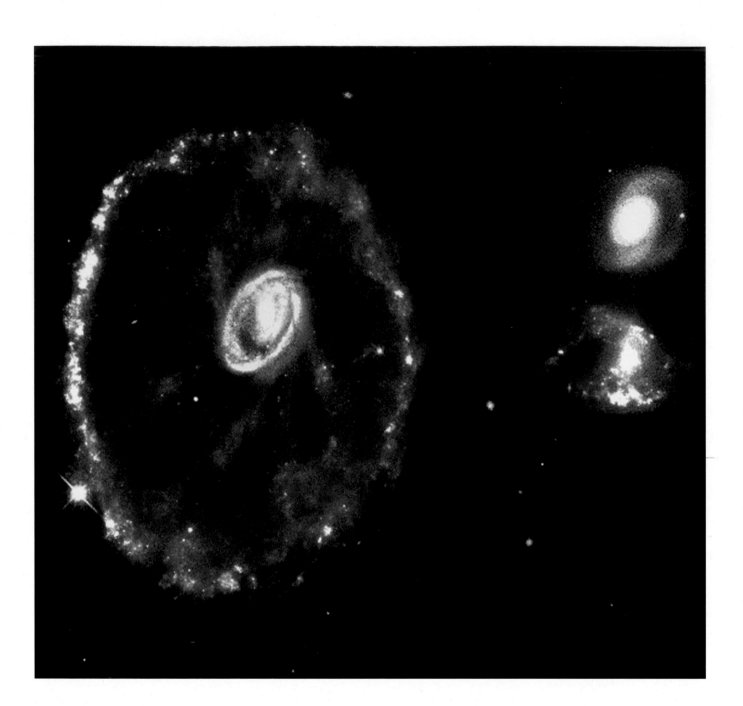

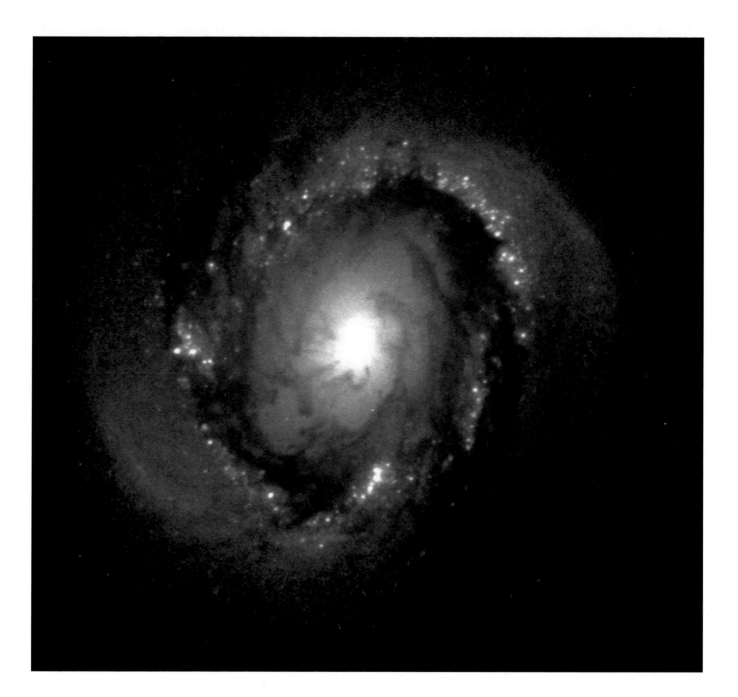

GALAXIES IN CHANGE

Testifying to a head-on collision with a past galactic intruder, the Cartwheel Galaxy (opposite) bears spokelike structures linking its central region to a striking outer ring. Like a rock tossed into water, the collision spawned a circular ripple of energy that plowed dust and gas before it, leaving in its wake cosmic wreckage: exploding supernovae, immense loops and bubbles, and bright blue knots that signify gigantic clusters of newborn stars.

Galaxy NGC 4314 (above), a type known as a barred spiral, is billions of years old. But its unusual, bright inner ring shows clusters of infant stars, indicating that its appearance has changed markedly in the past five million years. Is this old galaxy up to new tricks? Interestingly, its inner region appears much like a miniature spiral galaxy itself, complete with dust lanes and spiral arms—although it measures just a few thousand light-years across.

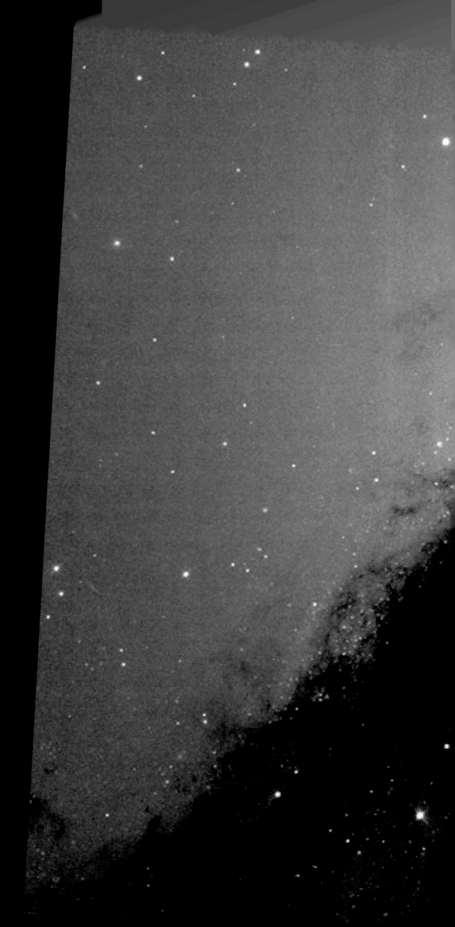

GALACTIC CANNIBALISM

Center of a maelstrom of activity,
a massive black hole hidden in
nearby galaxy Centaurus A feeds
on a smaller spiral galaxy (bright
spot at five o'clock), fueling its
already huge gravitational pull. An
immense lane of dark dust girdles
the larger galaxy and its orange
disk of hot gas. The disk is
perpendicular to the plane of
dust—while the black hole lies at
an angle to both. Dusty streamers
filigree glowing gases and blue
clusters of new stars.

At ten million light-years away,
Centaurus A is the nearest active
galactic nucleus known. Its black
hole may contain the material of a
billion suns—packed into a space
about the size of our solar system.

NEXT-DOOR GALAXY

FOLLOWING PAGES] Two million
light-years away, the glorious
Andromeda Galaxy, our nearest
major galactic neighbor, graces
the night sky. Close-ups from
the Hubble Space Telescope
reveal a double nucleus, possibly
indicating that this galaxy previously cannibalized a smaller one

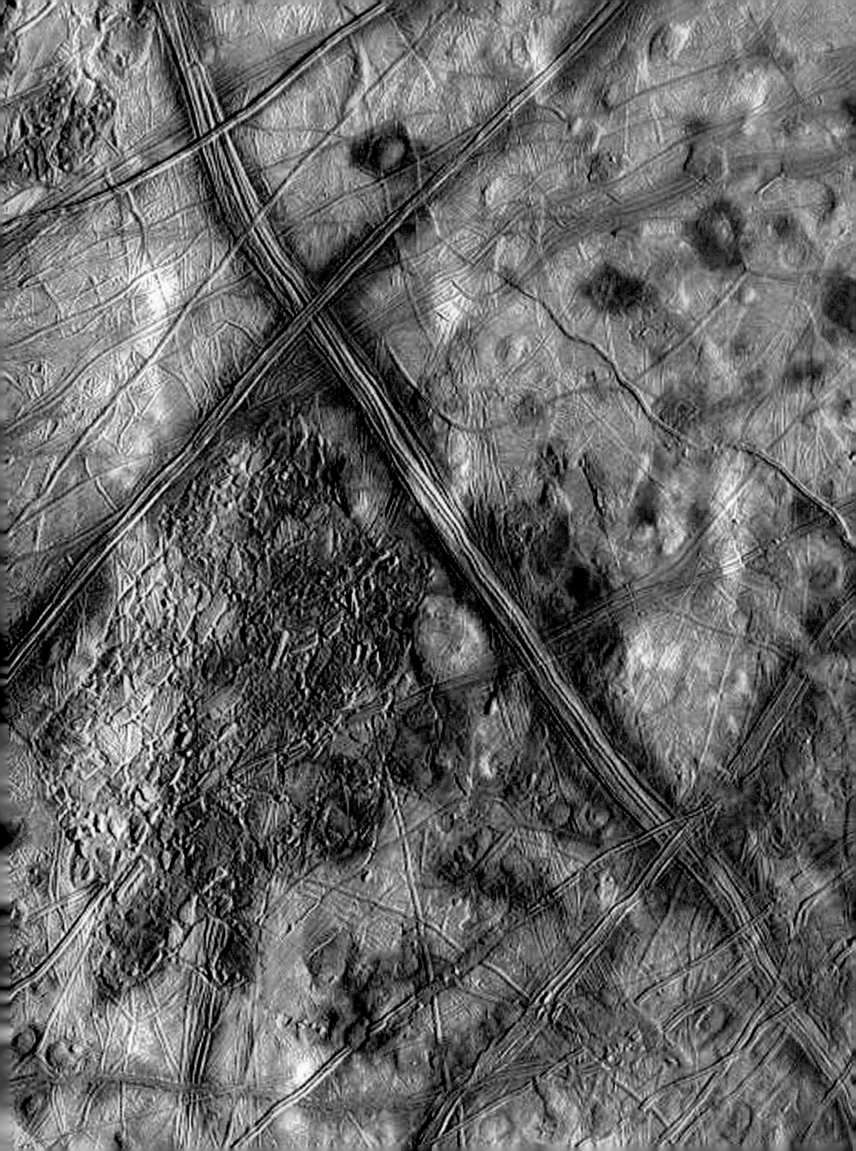

MYSTERIES OF THE COSMOS

HOW DID THE UNIVERSE BEGIN? HOW WILL IT END? IS THERE LIFE, MAYBE EVEN intelligent life, out there somewhere? The things we don't know about the universe include some of the most fundamental questions we can ask. It is to these sorts of deep, almost mythic questions that the next generation of astronomers will turn.

IS THERE LIFE OUT THERE?

Generations of science-fiction movies have conditioned us to consider bug-eyed monsters, ETs, and other rather sophisticated forms as typical examples of life outside the Earth. The reality, however, is that finding any kind of life at all, even something as simple as bacteria, would be one of the most exciting discoveries ever made.

The reason for this is simple: Every living thing on Earth, from the simplest microbe to the most complex organism, can be thought of as the result of a single experiment. Scientists believe we are all descended from that first cell that appeared in the Earth's ocean billions of years ago. This means that everything we know about living systems comes from that single cell and its descendants. One of the basic rules of science is that it is very difficult to come to a general conclusion on the basis of any single experiment. So when the only planet that we knew about in any detail was Earth, we had a much different picture of what it was than we do now. Seeing our home as just one planet among many gives us different insights into how it operates. In the same way, should we find life anywhere else in our solar system, we will gain enormous perspective on the nature of life in general, not just the nature of life on Earth.

If we do find evidence of extraterrestrial life some day—on Europa, perhaps, or Mars— then we will know there is a good chance that life is a fairly ordinary component of the

A MOON WITH WATER?

Liquid water may lurk beneath Europa's crustal plates and linear ridges, raising speculation that this Jovian moon could also harbor life. Blue signifies old ice; red regions may indicate more recent internal geological activity. White blotches are bright ejecta from an impact crater 600 miles away. This Galileo composite spans an area of about 120 by 150 miles.

universe as a whole. We already know life began on Earth when conditions were right, and any evidence that it arose elsewhere under different circumstances would be strong evidence that life takes root whenever and wherever it can, that it is more versatile than we ever imagined. If, on the other hand, we find that life appeared on Earth and nowhere else in the solar system, that might indicate that life requires more than what we now understand to be the right conditions to develop. There might be some additional necessities—for example, the presence of a large moon capable of raising tides on its home planet. If this is indeed the case, life could well be rare among the stars.

Another thing we would gain if we were to find living things in the Martian permafrost or the subsurface oceans of Europa would be insight into the methodologies of life. Right now, we know that all living things on Earth share the same genetic coding system and use the same basic DNA / RNA mechanisms to function. But because we have only this one system, we cannot tell whether it is a chance development of a particular chemistry within a particular cell billions of years ago, or whether it is a feature common to all forms of life. Are there reasons why this particular biochemistry has come to dominate Earth? Is there something we don't yet understand about DNA that makes it inevitable in living systems? If we had a bit of pond scum from Europa, we would know whether it contained our kind of DNA or RNA, or if its chemistry was something entirely different. Either answer is important, because either would change the way we think about living things.

TODAY, THE CONSENSUS WITHIN THE SCIENTIFIC COMMUNITY SEEMS TO BE that we eventually will find not only life in other parts of the galaxy but also intelligent and technologically advanced life. I have to say that I disagree. While I believe we will find other forms of life in other solar systems (if not in our own), I also feel it is extremely unlikely that a large number of advanced technological civilizations are out there, waiting to be discovered. The most succinct support for my view comes from Nobel laureate Enrico Fermi, the man who ran the first nuclear reaction ever controlled by human beings. Confronted at a 1950 luncheon with scientific arguments for the ubiquity of technologically advanced civilizations, he supposedly said, "So where is everybody?"

This so-called Fermi argument embodies a simple logic. Human beings have had modern science only a few hundred years, and already we have moved into space. It is not hard to imagine that in a few hundred more years we will be a starfaring people, colonizing other systems. (Most starship schemes are multi-generational, given the immense distances, which dictate voyages at least hundreds of years long just to get from one star system to another.) Fermi's argument maintains that it is extremely unlikely that many other civilizations discovered science at exactly the same time we did. Had they acquired science even a thousand years earlier than we, they now could be so much more advanced that they would already be colonizing our solar system.

If, on the other hand, they are a thousand years behind us, we will likely arrive at their home planet before they even begin sending us radio signals. Technological advances build upon each other, increasing technological abilities faster than most people anticipate. Imagine, for example, how astounded even a great scientist like Isaac Newton would be by our current global communication system, were he alive today.

Where are those highly advanced extraterrestrial civilizations so dear to the hearts of science-fiction writers? Their existence, I believe, is far from a foregone conclusion.

We have known since the 1920s, when Edwin Hubble made measurements of distant galaxies, that the universe is expanding. Other galaxies are moving away from us, with more distant galaxies moving away faster. The ultimate fate of the universe hinges upon whether or not that expansion will ever end. Physicists currently recognize three possible scenarios. In the first, called an open universe, expansion simply continues forever. Eventually the stars burn out and the universe becomes a cold, expanding sea of stellar cinders and isolated particles. A second possibility is that the expansion will stop and reverse itself. In this case, the Big Bang will give way to a Big Crunch, in which all matter in the universe will eventually return to something like its original, high-density state—perhaps to explode into another cycle at a later time. This is called a closed universe. Between the closed and open universes lies option three: a universe in which expansion slows down and comes to a stop after an infinite amount of time has passed. This is called a flat universe.

Which scenario is correct? Imagine a distant galaxy speeding away from Earth. In the conventional cosmological picture of the universe, the only force acting on that galaxy is the gravitational attraction of all other objects in the universe. Count up all the matter that exists, calculate the gravitational attraction it exerts on that particular galaxy, and it's possible to compute whether it will be slowed down enough to reverse its direction. By the beginning of the 1990s, astronomers had found only about 30 percent of the matter needed to "close" the universe, that is, to assure the inevitability of a Big Crunch. Their estimates included all known dark matter, so at that time, the fate of the universe seemed to revolve on whether or not anyone could find the missing 70 percent of matter in the cosmos.

Over the last decade, however, the logic of an expanding universe has dictated a new approach. When you look at light from a galaxy that is, say, ten billion light-years away, the light waves impinging on your retina actually left that galaxy ten billion years ago. You are seeing it not as it is but as it was at that time. Calculate the expansion rate of that galaxy from the light you receive now, and you have a measure of the expansion of the universe *as it was ten billion years ago*. This tells you how much expansion has slowed down or speeded up between then and now, and from this you can deduce the fate of the universe. In other words, rather than measuring matter itself, cosmologists now measure the gravitational force exerted by that matter, by measuring changes in the expansion of the universe.

The reason that they could not do this earlier is that it is very difficult to accurately gauge distances to galaxies that are billions of light-years away. Telescopes just cannot pick out individual stars in such galaxies, and so the Cepheid-variable standard candle used by Hubble and his successors doesn't help.

But now astronomers have developed a new kind of standard candle called the Type Ia supernova. This concerns a particular kind of double-star system, in which a normal star and a white dwarf circle each other. Over the course of time, the white dwarf pulls matter from its partner, creating a layer of material (mostly hydrogen) on its own surface. When that layer becomes sufficiently massive, tremendous pressures and temperatures are created, igniting nuclear reactions in the hydrogen and causing the star to "flare off." For a period of several months, the star shines brightly while this fuel is being burned.

The reason that Type Ia supernovae can serve as standard candles is that all white dwarfs in this type of system accumulate matter from their companion stars up to the point that their mass reaches about 1.44 times the mass of our sun. Then they explode. Because they explode

NEW STANDARD

Today's astronomical "standard candle," the bright spot fringing this galaxy is a Type Ia supernova, which exploded long ago. Because all Type Ia supernovae erupt with equal energy, we can calculate their distance according to their apparent brightness on Earth. Just as Cepheid variables became the standard for Edwin Hubble, these stellar remnants promise to expand our conception of the complexity and scale of our universe.

only when they reach this same critical mass, all Type Ia supernovae should be similar. Astronomers working with relatively nearby supernovae have found that by watching the way their core stars flare up and die out, they can deduce how much light they actually emit.

This means that when they see a Type Ia supernova go off in a distant galaxy, they can compute the amount of light being emitted and compare it to the amount of light actually received. From this comparison they then deduce the distance to the galaxy. So far, well over a hundred Type Ia supernovae have been spotted in distant galaxies, each one providing a determination of its distance from Earth. When astronomers compare current speeds of these distant galaxies to the speeds at which they are receding from us, an astonishing fact emerges. The Hubble expansion does not slow down at all—in fact, against all expectations, the rate of expansion has increased over the last 10 or 12 billion years.

The only way we can resolve this is to accept that there is more going on here than simple gravity. Newton's gravitational laws just don't explain what we observe. Therefore, there must be another force acting at large distances in the universe—a kind of antigravity that tends to push galaxies apart. Believe it, sci-fi buffs: Antigravity actually seems to exist. In fact, Albert Einstein proposed such a force back in the 1920s, describing it in terms of what he called a "cosmological constant," but quickly abandoned it. Today's emerging view is that the fate of the universe results from a competition in which the inward force of gravity is countered in some way by the outward push associated with this mysterious constant. On one level, this means that the universe will expand forever. On another level, it means that over the next several decades astronomers are going to be very busy exploiting their newfound ability to determine distances by means of supernovae. This, in turn, should give us not just a general notion of the fate of the universe but also a detailed picture of how it will actually end.

I hasten to add that the distances involved are so huge as to be almost beyond human comprehension. Einstein's cosmological constant is something that acts over distances of billions of light-years. It isn't going to produce any major changes in our current view of the solar system or even the Milky Way galaxy. Nevertheless, it is a landmark discovery, since a primary goal of astronomy is to completely understand the universe we live in. Clearly, this effect will help us fill in that understanding.

HOW DID THE UNIVERSE BEGIN?

Just as understanding the end of the universe is a crucial part of astronomy, so too is understanding the beginning. Everything we currently know indicates that the universe began in an extremely hot and dense state billions of years ago. One measure of our understanding of it is to ask how far back toward the beginning we can go and still explain the state of the universe. Ultimately, such a trek will bring us to the question of whether we can explain creation. The scientific and philosophical ramifications of being able to ask—and possibly answer—such a question are enormous. Indeed, they explain the drive of modern scientists to probe ever back in time, toward the moment of creation itself.

To describe the earliest cosmological times, it is necessary to understand how matter behaves at extremely high energies. High temperatures cause the constituents of matter to move at high speeds and collide violently.

We know that when we break down the nuclei of atoms, we find elementary particles—protons, electrons, neutrons, and such—and when we break down these particles, we find that they are made of things even more elementary, called quarks. By using high-speed particle accelerators such as those at the European Laboratory for Nuclear Research (CERN) in Geneva, Switzerland, or the Fermi National Accelerator Laboratory near Chicago, we can smash different particles into each other head on, at velocities near the speed of light. For a brief fraction of a second, in a volume the size of a proton, these collisions create temperatures similar to those we believe existed when the universe was only one ten-billionth of a second old. In other words, we can produce conditions in our labs that have not existed in the universe since that time, and our theories explain those interactions very well. What this means is that we can now trace the origin of the universe in both the experimental and theoretical realms back to one ten-billionth of a second from the beginning.

We also have theories (although no experiments yet) that can take us back much farther still—to 10^{-35} seconds (that is, a decimal point followed by 34 zeros and a one) after creation. Although we cannot reproduce in the lab the kinds of temperatures we believe existed then, we can show that many features of our universe are explained by current theories. Thus, we can say that this tiny fraction of a second marks the current theoretical frontier of our knowledge of the Big Bang.

In addition, scientists are speculating on theories—so-called "theories of everything"—that would take us even farther back. Over the next several decades, progress in this field will most likely come primarily from researchers who develop logical ways to explain the most fundamental constituents of matter and their interactions. How we understand the origin of the universe will, in the end, depend on what shape those theories take. It is not too much to expect that, when we have such theories, we will be able to comprehend how a universe as full of matter as the one we live in arose from a vacuum. Some day, we even may be able to answer the ultimate question: Why does the universe exist at all?

Of course, we *Homo sapiens* are descended from primates that came down from the trees and began to walk upright only some four million years ago. The human brain that evolved gave our ancestors a better chance of surviving, even in hostile environments, and it has done the same for us. And now that brain is able to conceive, speculate, and discuss what happened during the creation of the entire universe some 10 or 12 billion years ago, as well as what fate that universe will face hundreds of billions of years in the future. Certainly the story of the cosmos is an awesome one. But so is the story of our species. ●

PAST AND FUTURE

[FOLLOWING PAGES]

Whipped by fierce stellar winds, hot clumps of gas race outward from the raging fireball called WR124, an extremely rare, short-lived, and superhot class of star that is undergoing a violent transition, probably on its way to becoming a white dwarf. Its surrounding nebula is thought to be no older than 10,000 years, so young that it has not yet slammed into the interstellar gases beyond, as occurs with more mature nebulae.

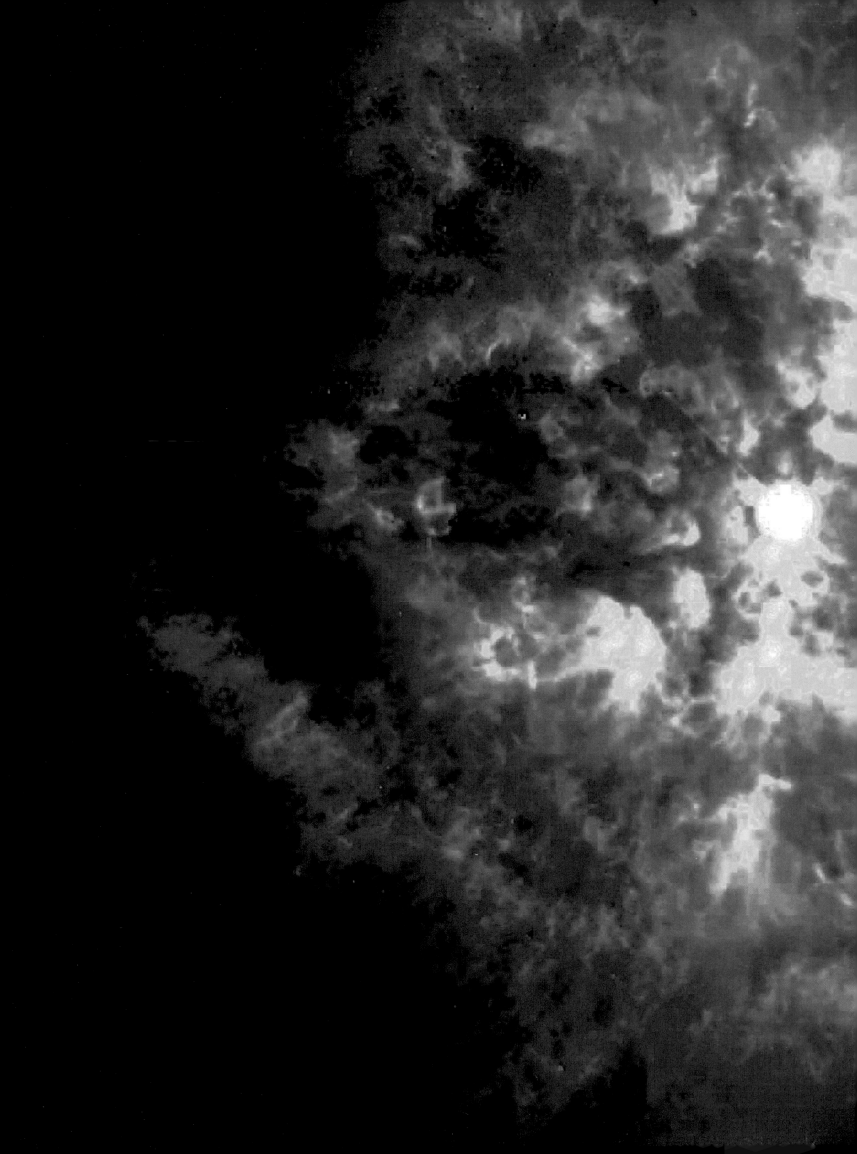